D0312938

Mobile DJ Handbook

Second Edition

How to Start and Run a Profitable Mobile Disc Jockey Service

Stacy Zemon

Focal Press

An Imprint of Elsevier

Focal Press
An Imprint of Elsevier

ISBN 0-240-80489-9

British Library Cataloguing-in-Publication Data
A catalogue record for this book is available from the British Library.

The publisher offers special discounts on bulk orders of this book. For information, please contact:

Manager of Special Sales
Elsevier
200 Wheeler Road
Burlington, MA 01803
Tel: 781-313-4700
Fax: 781-313-4882

For information on all Focal Press publications available, contact our World Wide Web home page at: http://www.focalpress.com

10 9 8 7 6 5 4 3 2

Printed in the United States of America

This book is dedicated to all mobile disc jockeys
from the industry's beginning,
who have contributed so much
to the evolution of the business.

If you want to lift yourself up, lift up someone else.
                              BOOKER T. WASHINGTON

*When you are inspired by some great purpose, some
extraordinary project, all your thoughts break their
bonds; your mind transcends limitations, your
consciousness expands in every direction, and you find
yourself in a new great and wonderful world. Dormant
forces, faculties' talents become alive, and you discover
yourself to be a greater person by far than you ever
dreamed yourself to be.*
                              PATANJALI

Table of Contents

        Sound Reinforcement Terms   145

Chapter 11.   How's Your Form?   171
        Sample Client Contract   173
        Sample Wedding Party Introduction Sheet   176
        Sample Employment Agreement   178

Chapter 12.   Records of Success   191
        Interviews with America's Top DJs   191

        Appendix   215
        Index   229

Foreword

By Jim Robinson, *Promo Only*, Co-Founder/President
Electro-Magic Productions, President/Owner

My good friend, mobile DJ icon, Bernie Howard, has often accused me of owning a crystal ball. I don't know about that, but I have had some pretty good seats to see firsthand, the changes in the industry over the past 25 years. Please allow me to share my story.

I started back in 1976 as many of us did, as a "bedroom" DJ hungry for 12-inch versions of dance songs. Back then there were records, reel-to-reel, or the highly uncool 8-track. I had all three.

A few years later I got my first music-related job in Disney World's tech services division, the department that handled all their sound needs. I jumped at the chance to learn how to put together concert-size sound systems for artists like *KC & the Sunshine Band*, *Kool & the Gang*, *Peaches & Herb*, and many other bands who were popular at the time.

In those days I spent most of my time lugging huge amps and speakers. It occurred to me that I could benefit from using what I learned—and that if I did gigs through my own company, I could probably use my home-stereo equipment costs as a tax write-off. I would rent the big stuff when I needed it. Thus, Electro-Magic Productions was born. Little did I know that it would be the start of a lifelong career.

In 1981 I transferred to engineering, where I set up sound systems that piped ambient music to Epcot Center on high-speed tape. In this job I learned from the world's best sound engineers for those types of systems. Those guys were very picky and the quality had to be exactly right.

Not long after I started college where, naturally, I was a DJ at fraternity and sorority parties. I discovered that I was good at and loved entertaining people. So, in 1986, I left my full-time job to concentrate on school. My intentions were to make money as a mobile all through college, then graduate and get a "real" job. However, that's not how my story evolved.

To increase my DJ business while still in school I got a job as a radio personality for a Florida-based, Top-40 station. I also handled all of their special events and pool parties.

We had consumer compact disc players but no CDs, so we had to rely on using records. Out in the hot Florida sun I would watch the vinyl warp before my very eyes. Out of necessity I borrowed two cart machines and carts[1] from the radio station and used them. I was lucky to be able to borrow them for outdoor events because they were very expensive. Thankfully, when I worked at clubs, I could beat mix using my beloved vinyl.

About a year later, CD players for disc jockeys were made available. I saw one for the first time at The Lighting Dimension International (LDI) convention in Orlando, Florida, which was the predecessor to DJ-specific trade shows (which sprang up later). The compact disc players for DJs were large, didn't work very well, and were expensive. I opted instead for two portable players. After 3 months I had gone through three players because they were dropped during set up or take down at gigs. My solution? I made my own "professional" CD player by drilling holes in the case of a consumer model player and rack mounting it.

During the fall of 1991, the *DJ Times* trade show came to my area. I think it was the second all-DJ convention ever held. Johnny James of *Star DJs* in New Jersey brought more than 20 of his over 100 mobiles to the conference. I believe he wanted to show everyone at the convention that being a disc jockey could be a lucrative career. I was impressed and motivated by what I saw and heard.

---

[1]Carts looked like 8-track tapes with reel-to-reel tape inside them and had only one stereo track. Radio stations used them to play music on the air because they were high quality and automatically recued quickly when finished playing. Carts were typically 4 minutes in length.

One year later, in 1992, Denon introduced a piece of equipment that many of us today could not imagine working without—compact disc players with two players in one unit. These were the next evolutionary leap from the Techniques 1200 turntables and featured pitch control, instant start, cueing capability, and remote controls. They were even reasonably priced. Unfortunately, there were few music CDs on the market to play in them. I discussed this void with Pete Werner, a friend who became my business partner. Out of those conversations, *Promo Only*, the monthly CD music service was born. At the time, Pete was a reporting DJ for *Billboard* magazine. He knew the music and I knew the technology, so we were a good team. We believed that offering the current club hits on one compact disc was an idea whose time had come. Since that time, the needs of our subscribers have dictated all of our decisions in this area.

Starting our business in 1992 was a pretty bold move because the record companies were still undecided about their commitment to analog and digital formats for club music. The industry struggled to find direction, a quandary that hasn't changed much in the past 10 years. Many club songs and mixes are still only available promotionally on vinyl, while other mixes or edits are only on promotional CD, and are still not for sale. I'm sure many DJs have felt frustrated when trying to deal with the music industry's lack of decisiveness.

Since those early days our company has expanded into the Canadian and European marketplaces. We have added many series of music on compact disc, geared toward different types of DJs and audiences.

In 1998, digital video disc players came onto the scene and have been widely accepted by the general public. DVD is the next generation of optical disc storage technology and offers significant quality superiority over VHS. Once again, necessity was the mother of invention, so *Promo Only* began offering a series of DVD music videos in 2000. Although currently VJs use consumer players, my crystal ball tells me that it's only a matter of time before we see DVD players made for VJ/DJs.

Twenty years after the introduction of the first professional compact disc player, Pioneer innovated a unit that simulates working with vinyl on turntables using a CD. This technology has birthed a whole new breed of disc jockey; one who creates original music and uses this "digital turntable" more like an instrument than a piece of gear. This new generation of DJs has changed the game and raised the ceiling for all of us. I predict that before the end of the decade, we will see these creative pioneers revered as highly as vinyl-spinning DJs have been. An outcome will be that the Grammy Award category for "Best Remixer" [2] will then finally be televised.

It's nearly 2003 as I write this Foreword. The choice of technologies, media, and manufacturers continues to grow. Equipment is much more sophisticated than it used to be and we can loop, sample, and overlay to our heart's content. So, how do DJs figure out what they need and what to buy?

Using common sense and comparing the economic feasibility of each choice is a good place to start. Computers, compact- and mini-disc players and discs, turntables, MP3 (and all the rest) have their advantages and disadvantages. Bear in mind, with whatever you choose, clients don't care what you use as long as you give them the music, quality, and experience they seek.

To put a spin on what William Shakespeare once said, To be free or not to be free? That is the question. As we go to print, record labels, defended by the RIAA, are seriously threatened by the ever-growing free-music philosophy. Sure, the Internet has made music more accessible to the masses, who generally are apathetic to intellectual property concerns but—is stealing right? Can the unlimited free download go on forever? Can you trust it? Will the new copy protection on CDs work? Will all music ever be available legally on the Internet? Let me see, where's my crystal ball?

The future about these questions is fuzzy but there are some things that are very clear. It's an exciting time to be a mobile disc jockey entertainer. We can now provide pure digital quality audio and video as well as spectacular lighting effects at an affordable price for our customers. With today's high-tech equipment and learning opportunities available through books, trade shows, Web sites, and professional DJ stores, we are limited only by our own imaginations. The mobile disc jockey business has matured into a profession that is recognized as a legitimate occupation and industry with its own labor code.

New improvements will continue to come. One thing that will never change, though, is what is required to be successful as a

---

[2] The Best Remixer Award is presented to the person who is selected to have done the best remix of a song. To date, A DJ/producer has won in this category every time.

mobile disc jockey entertainer: talent, knowledge, determination, and dedication. Use the information in this book to help you—and go ahead—write your own story. I'll be looking into my crystal ball, watching *you* take *us* to the next level, and I'll be smiling all the while. Good luck!

Acknowledgments

First, last, and always, I want to acknowledge and thank God, the Source of life without whom neither this book nor I would have been written.

I also extend my gratitude to the wonderful team at Focal Press. Thank you to: Marie Lee, Diane Wurzel, Bruce Siebert, and the publishing board for saying "yes" to me—not just once but twice—and for all of the personal and marketing support you have given me and my books.

My humble admiration and sincere appreciation goes out to: Jim Robinson, President of *Promo Only* for writing the Foreword to this book and for his contribution to Chapter 5, Your Music Selection; to DJ icon Bernie Howard Fryman for his ongoing support and huge contribution to Chapter 4, Getting Geared Up; and to DJ/Webwizard, Sid Vanderpool, for his invaluable input. Thank you all—BIG TIME!

With great respect I'd like to acknowledge some special DJ folk who are the absolute tops in the business: Keith Alan, Mike Alexander, Mark Ashe, Randy Bartlett, Paul Binder, Mike Buonaccorso, Lisa Capitenelli, Bob Deyoe, Jennifer Deyoe, Brian Doyle, Frank Garcia, Peter Goldsmith, Jeffrey Greene, Roxanna Greene, Richie Hart, Kevin Howard, Gary Kassor, William Kozma, Chuck Lehnard, Ray Mardo, Kem Matthews, Bobby Morgenstein, Brian O'Connor, Ed Price, John Roberts, J. R. Silva, Jeff Stiles, Mark Thomas, Randi Rae Treibitz, Jim Tremayne, Tony Valentine, Gerald Webb, and Adam Weitz. Thank you for all that you have shared and the opportunities that you have provided.

For those people in my personal life who have given so much that matters, I am truly thankful to: Pam Belknap, Beth Diamond,

Rob Feiner, Gene Goodman, Gloria Goodman, Russell Goodman, Rosemary Hargreaves, Hilarie Jones, Carolyn Kydd, Jacob Lee, Krishna Sondhi, Jennifer Suprenant, Wendy Trager, Mike Williamson, John Wolfson, Joan Wurzel, Robert Zemon, and Tom Zemon.

My deep gratitude is offered up to the spiritual teachers who have profoundly made a difference to me: Shayna Appel, Jane Cantrell, Joyce Cohen, Milbrey Ewing, Gordon Feller, Lisa Hupfer, Patricia Ireland, Patricia Johnson, Landmark Education, Robin Lunn, Shirley MacLaine, Ian Mayo-Smith, Gerianne Robinson, Gene Roddenberry, Robert Roth, William Spady, Diane Sullivan, Louis Turner, Loretta Vasso, and Neale Donald Walsch. Thank you all for lighting my way and for teaching me how to do the same for others.

Last, but certainly not least, I wish to acknowledge and recognize all of the colleagues, mentors, and friends, named and unnamed, who have taught me so much about entertaining, business, and life along the journey—and to you, dear reader, for demonstrating the wisdom to buy this book.

Introduction

With the increasing popularity of DJ-oriented Web sites and books, today there are more resources than ever to assist you in becoming a professional mobile disc jockey entertainer and in starting a profitable DJ service.

Because of the popularity of my first book, my publisher and I felt there was a need for a second edition that would incorporate all of the technological, as well as the many other, changes that have taken place in our industry. The pages that follow compile a current and comprehensive resource guide that will provide you with a solid foundation in the important elements needed for success in the mobile DJ business.

The Mobile DJ Handbook: How to Start and Run a Profitable Mobile Disc Jockey Service is the international #1 best seller that has been read by more DJs worldwide than any other book on the subject. Written for both newcomers and experienced professionals, it is the most comprehensive reference guide ever published for achieving success as the owner–operator of a mobile disc jockey company. It contains information on becoming a professional, securing bookings, buying equipment and music, and running party dances, contests, and games. It also includes details on how DJs can market and sell their services, plus, a wealth of specific ideas that will help them develop and expand their businesses.

For the beginner, this book will teach all of the elements involved to become a professional disc jockey entertainer. For the seasoned pro, *The Mobile DJ Handbook* will enable you to achieve great success and prosperity as the owner of your own mobile disc jockey service. I have written what I consider to be a blueprint to assist you in realizing and achieving these goals. This book offers advice on everything you need to know to gain the personal entertainment and business skills required to successfully work in this exciting field and to start your own company. Included are details on successfully selling your services, plus a wealth of specific ideas that will help you develop your business and to achieve personal prosperity.

The Mobile DJ Handbook is filled with practical tips and creative strategies and is written in simple, straightforward language. The beginner may find learning all of the details of this business to be overwhelming. This book was written to make the path easier. Getting the most out of this material requires deep involvement

Figure FM.1 Author Stacy Zemon's first microphone. Even as a baby, she knew how to shake, "rattle," and roll!

with it. When the ideas set forth are put into practice, the results can be phenomenal. The content of this handbook is the culmination of feedback from the industry's most successful DJs, as well as my own extensive experience.

Please let me know how the ideas in this book serve you by writing to me at djstacyz@aol.com. Read. Learn. Do. Grow. Share. And just one more thing, have FUN in the process! I challenge each of you to better your best . . . to go forth and become a profitable professional . . . to even be a DJ revolutionary. You are the future of the business and the one who will lead us on to the next evolutionary step! Remember, learning is a cumulative process that continues throughout your career. enJOY the process.

About the Author

Why is Stacy Zemon a leading authority on the mobile disc jockey entertainment industry? Well, for starters, she has nearly 25 years of experience working as a mobile, club, Karaoke, and radio DJ. She also wrote the world's top-selling book on the mobile DJ business. What's more, she pioneered the nation's most innovative disc jockey service. Whatever the harmonious blend of reasons, Zemon's career has produced an impressive record of success.

Stacy Zemon's interest in music and entertaining began as a teenager in Connecticut, and has never stopped! At age 13 she created an in-house radio station using the initials of her name as call letters on the now archaic equipment that was available in the early 1970s.

During the summers of 1977 and 1978, Stacy worked as a member of the social staff of Tamiment, a major resort in the Pocono Mountains of Pennsylvania. There she ran contests and games and taught participation dances to vacationers. By 1979, she was working in commercial radio in the Nashville, Tennessee, area. At this time she began DJ-ing in clubs to supplement her income.

After a few years, Stacy moved to Philadelphia where she continued her radio and mobile work in Philly and South Jersey. She quickly added being a Karaoke entertainer to her list of performance abilities. Today she can be heard on the air in the Pioneer Valley of Western Massachusetts on WPVQ-FM 95.3.

Over the years her radio work has included being an announcer, music director, program director, and operations manager in Top-40, adult contemporary, contemporary hit radio, and country music formats. She has worked for four mobile DJ companies, has started two of her own, and consults to several others across the country.

A groundbreaking concept was originated by Zemon after completing *The Mobile DJ Handbook*. She partnered with a national radio broadcasting company to form the first corporation ever to directly link mobile disc jockey services to radio stations. As resident and hands-on manager, Zemon began a multi-system operation that was the mobile DJ entertainment division of seven stations located in Connecticut and Massachusetts.

The fast track to success was assured because no other area DJ business could compete with the company's marketing ability or radio station affiliations. The mobile DJ services were primarily promoted through commercials aired on each station. Equipped with a team of 23 DJ personalities, Zemon's crew was booked regularly throughout Connecticut, Massachusetts, and New York. Her company offered Mobile DJs, radio personalities, light shows, high-energy hip-hop dancers, party props, prize packs, photography, video, and more. It utilized the best quality sound and lighting systems and a giant CD music library covering virtually every genre of music from the 1940s to the latest hits.

Clients responded with great enthusiasm to the one-stop shopping concept and to the quality and professionalism that was delivered. Within its first year of operation, the company became the most requested and well-known mobile DJ service in New England. Zemon eventually sold her business interest to her partner, who was then bought out by another radio broadcasting company.

Zemon has shared her consummate knowledge of the mobile disc jockey industry with her peers both as a writer for *DJ Times* magazine and on site at their International DJ Expo, held annually in Atlantic City, New Jersey. She has become a familiar face to conference attendees as the facilitator of leadership and marketing seminars and as a judge at Disc Jockey of the Year competitions.

Her love for entertaining party goers, music, fun, and inspiring other disc jockeys is what continues to motivate Zemon to be the best. Currently, she entertains at select VIP functions, produces large-scale events, is a consultant to mobile music services, and is involved in other DJ-related entrepreneurial endeavors. Further information on Zemon can be found at www.themobiledjhandbook.com, the on-line DJ community WebSite.

As an entertainer, author, and innovator, Stacy Zemon's creative wizardry has catapulted her into the role of being the Ray Kroc of the mobile disc jockey industry.

Figure FM.2 Stacy Zemon, Northampton, Massachussetts.

Do not wait; the time will never be "just right." Start where you stand, and work with whatever tools you may have at your command, and better tools will be found as you go along.

NAPOLEON HILL

Your 21st Century Mobile DJ Service

Honesty, integrity, and communication are the three most important words to remember when starting your DJ business. By being honest with yourself and with your clients you will create an atmosphere of trust. By demonstrating integrity you will earn respect. Clear communication will help you to avoid misunderstandings. Be willing to ask questions when you are unsure and to inform others when an unforeseen problem arises.

MARK THOMAS, OWNER
Awesome Entertainment
Pasadena, California
2001/2002 American Disc Jockey Association
President

Creating Prosperity and Success

Success can be defined as the continued expansion of happiness and the progressive realization of worthy goals. Successful people understand that there is no such thing as failure, only mistakes from which they can learn. They are persistent, dedicated risk takers who have an unwavering belief in themselves and what they are doing. They aim high, learn quickly from their experiences, are action oriented, and love challenges.

To attain success, you must learn to focus on the journey and not just on the destination. Before you set out on your journey, you ought to know something about where you want to go and when you expect to arrive.

A necessary ingredient in any formula for success is vision—your underlying, driving, desire-filled concept of what you value most in life. To convert a vision into a goal, break it down into a workable action plan of daily, weekly, and monthly steps. First, write out a vision statement. Detail for yourself what you want from your business and what will motivate you to take action to achieve it. You can be as specific as you want and should include questions relevant to all of your hopes and dreams. If the sky were the limit, what would you want? Write a list of your emotional, intellectual, physical, spiritual, and monetary goals. A mission statement reflects your vision of your ideal company.

When your list is complete, eliminate all of the items you wrote down because you thought you *should*, not because you really want or need them. Ask yourself if you are willing to make a plan to attain or achieve each of the remaining items on your list. To be successful you must act as if you are already a success. You must believe that you can succeed. If this is not your perception, you may sabotage your own efforts. At first, acting like a successful person may feel as though you are simply an actor playing a part. Eventually, the "acting" will translate into "being." Demonstrate an attitude of success in every area of your life, such as your manner of dress, the way in which you speak with people, and how you conduct yourself and your affairs.

Be enthusiastic! Verbalize your excitement about your business to everyone you speak with. Stay in contact with mentors who have what you want and are willing to talk with you about how they got it. Whatever your level of success, be willing to share it with others. To keep something, you must be willing to give it away!

Learn to attain and keep a positive attitude by reading motivational books or by listening to tapes of positive affirmations. *Positive thinking* is seeing something good in every situation. Successful people have the ability to focus on the positive and have faith that their goals will manifest themselves. Faith is very different from hope. *Faith* is the inner knowledge of things unseen. Your goals can be manifested through the use of prayer, meditation, affirmations, and action. Here are some ideas to assist you:

- As often as possible, focus your attention on exactly what you want from your business and your life.
- Commit to your success.
- Do the work required.
- Remove all obstacles that stand in your way.
- Visualize having what you desire.
- Believe you deserve what you are visualizing.
- Act as if you already have these things.
- Love your work and constantly strive for improvement.
- Be honest and keep your commitments.
- Have faith and let go of fear and resentments.
- Be patient and persistent.
- Find ways to "recharge your batteries."
- Help others to have what you want for yourself.
- Be truly thankful for what you receive.

If you love being a mobile DJ and focus on service, financial rewards will naturally follow. The focus must be on the service, not on the money. Financial rewards are always the secondary outcome of serving others well.

Becoming a Professional

Professional mobile DJs come from all walks of life. They are men and women from every ethnic, racial, and socioeconomic background and age group. Mobile DJs travel to various functions and locations to be emcees and to entertain guests through music and motivating speech. The DJ, or the company for which he or she works, owns the equipment and music brought to the event. Experienced mobile disc jockeys are highly skilled professionals. Those who own their own businesses have made a major investment of their finances, time, and reputation. It is necessary to have the proper experience and resources in place before starting your own business.

In the mobile DJ business your performance is your product. When your product is excellent, you will receive referrals that lead to bookings. Consistent bookings will create the cash flow that is necessary to promote, improve, and expand your business.

The Mobile DJ Handbook attempts to illustrate the aspects of the profession that can be taught. As a general rule, the best mobile DJs are outgoing individuals with a high energy level and a commanding presence. They usually have pleasant speaking voices and are able to think quickly, communicate clearly, and motivate others. They enjoy a wide variety of music and have a passion for entertaining a crowd. These people desire to continue to learn and grow as performers.

Mobile DJs who successfully run their own companies are savvy businesspeople with strong sales abilities. Their businesses are adequately capitalized to finance the marketing and advertising necessary for promotion of the company. They consistently give excellent performances and are highly customer-service oriented.

Not all DJs have the ability to perform equally well at different types of occasions. A high school prom calls for an entirely different kind of music and entertainment style than does a dance for 35- to 60-year-olds at a country club. Ask yourself the following questions: Can I relate equally well to both audiences? Do I enjoy alternative and rap as much as Motown and big band? Your answers to these kinds of questions will determine the types of occasions that you may choose as your specialties. It will also help you determine what kind of music library you will need. You may want to consider only performing at the types of events that will best utilize your talents. As your business develops and you hire other DJs, who may have different talents than yours, you can expand the types of functions performed by your company. Here are some ideas that can help you become a professional mobile disc jockey:

- Set short- and long-term goals. Review these goals daily and visualize yourself achieving them.
- Stay abreast of what is happening in the industry by reading trade publications and attending DJ conventions.
- Constantly strive to improve your entertainment and business skills by learning from successful people.
- Read and watch educational and motivational books and videos.
- Learn from and get involved with successful people through networking organizations.

Honest self-evaluation is crucial when planning for success. Ask yourself the following questions to help you determine if you have "the right stuff" to start your own business:

- Do I have strong leadership abilities?
- Do I consider myself a savvy businessperson?
- Do I possess excellent sales abilities?
- Do I like to make my own decisions?
- Do I enjoy competition?
- Am I self-motivated and self-disciplined?
- Do I plan ahead?
- Do I get along well with people?
- Am I willing to put in both the financial and the time commitments to run my own business?
- Do I have the physical stamina to handle the work required?
- Do I have the emotional strength to withstand the strain?
- Do I possess polished and well-rounded DJ entertainer and emcee skills?
- Am I thoroughly familiar with how to run mixing activities, participation dances, contests, wedding receptions, and Bar/Bat Mitzvahs?

There are several books on the market that contain a wealth of musical facts and historical information. Some of these books contain chart toppers and spotlight current events that took place the year a particular song hit the charts. Having this kind of information can be a treasure to a DJ who wants to keep the audience entertained with interesting anecdotes and facts. Check out the music section of your local bookstores and library. These books are also available from DJ-oriented catalogs and Web sites.

There are trade publications and organizations that produce yearly mobile disc jockey expos. The opportunity to learn and grow from attendance at these conventions is incredible. They offer workshops and seminars on a variety of topics, and provide an important chance to network with and learn from many professionals in the field.

"Newbies" versus "Bottom-Feeders"

In the mobile DJ business there is a world of difference between being a "Newbie" and being a "Bottom-Feeder." In both cases their rates may be discounted but their business conduct and operations are entirely different.

A "Newbie" is someone who is new to the DJ business and simply lacks the experience, references, and connections that benefit the higher priced, established mobile disc jockey. Newbies are willing to continuously reinvest their DJ revenues into professional improvement by adding and updating their equipment, music library, attire, and other tools of the trade. They keep their rates competitive and are always aware that rates can slowly rise as the value of their equipment, performance, and professional expertise increases.

Someone who is new to the mobile business and wishes to be a professional must be willing to learn and share new ideas, get to know the competition, and network with DJs from other geographic areas. Pros carry liability and property insurance, use signed contracts, invest in backup equipment, give their clients personalized attention, and always act with integrity. Over time, these DJs will begin to attract a more discriminating and educated clientele. Referrals from past clients and years of networking will be their main source of business.

Bottom-Feeders only reinvest monies when their equipment stops working. These DJs are not concerned with professionalism, improving their skills, or learning anything new. Their music collection is typically very limited and they obtain clients by underbidding the competition and offering the lowest price. Unfortunately, many uneducated consumers fall prey to the financially attractive sounding sales pitch and do not realize, until the day of their event, that they truly did not get a bargain.

Repeat clients are not the mainstay of the Bottom-Feeder. These DJs do not bother to carry insurance, have little backup equipment, if any, and run their business like a sideline. Some will even make promises to clients that they do not intend to keep. Often they will subcontract the job to an even less expensive and inexperienced disc jockey.

If you plan to be a Newbie, welcome to the fold. If you plan to be a Bottom-Feeder, please stay away from the mobile disc jockey profession. Your actions are detrimental to your hard-working, professional competition and damage the reputation of the industry.

DJ Schools and Apprenticeships

Operating a mobile DJ service can be a part-time job or a full-time career. Whether you are male or female, if you have never

Figure 1.1 Mark Thomas (aka DJ Peace), Owner, Awesome Entertainment, Pasadena, California. Courtesy Paul Antico, Creative Antics.

worked as a mobile disc jockey, you will need solid skills before starting your own business. There are two primary ways to gain these skills, either by attending a DJ school or by working as an apprentice at an established disc jockey service. There are numerous companies that offer training to those who want to work for them.

An apprentice learns by observing the various styles of experts and then puts into practice what he or she has learned. Find out the name of the most successful and respected DJ service in your area. Call the owner and explain that you would like to observe some of their best disc jockeys in action because you are interested in becoming a mobile DJ. Offer to be an unpaid assistant at weddings, Bar/Bat Mitzvahs, and other types of affairs.

Once you are at an event, pay close attention. Take lots of notes and ask every question you can think of, such as, "How does the mixing console work? How do you decide what music to play? How do you beat mix? How do you get the attendees dancing?" You will eventually develop your own style and your own way of doing things, but first it is necessary to learn the basics.

As a keen observer of an event, you will quickly see that it is the DJ's responsibility to ensure that everyone has fun. This is accomplished through developing a feel for what the crowd wants and giving it to them through a combination of music and patter. "Patter" is the talking the DJ does between songs. This friendly chitchat will get the audience to relax and at times make people laugh. Generally it will help the guests get into the spirit of the party.

When you are fully confident that you can duplicate what you have learned, ask the owner for a job. Do not rush this process. Demonstrate your great attitude by offering to improve your skills by coming into their office to practice on their equipment. You may have to be willing to do practically anything you are asked to get your first professional break. Getting to "yes" may take persistence and patience, but it's absolutely necessary to achieve your goal.

Most professional DJ entertainers agree that it takes a couple of years to gain enough experience to venture out on your own. In every profession it is necessary to gain the experience needed for success.

Establishing Your Business

As the owner of a DJ service, your business skills (including sales and marketing) should be equal to your skills as an entertainer. Your time and attention will be split between these two areas.

The first step before making any investment is to honestly assess your own strengths and weaknesses as a disc jockey and as a businessperson. It may be wise to retain another source of income until your disc jockey service can support your lifestyle. Prior to starting your business, there are some important questions to ask yourself: Am I professionally and financially ready to start my own DJ service? Do I possess the multiple skills needed to be the DJ entertainer I want to advertise myself as? How will I finance my business? (If you do not have several thousands of dollars to invest, you may have to take out a loan. You may also want to consider starting a business with a partner.)

Here is a checklist that will help you to determine accurately if you are sufficiently prepared to go into business for yourself:

- Will you, if necessary, lower your standard of living until your DJ business is firmly established?

- Do you have the business skills you will need to run a successful business?
- Have you ever worked at a DJ service similar to the one you want to start?
- If you discover that you don't have the skills you need, are you willing to delay your plans until you have acquired the necessary skills?
- Will your business serve an existing market in which demand exceeds supply?
- Will your business be competitive pertaining to its quality, selection, price, and location?
- Have you identified your customers?
- Do you understand customer needs and desires?
- Have you chosen a name for your business?
- Have you chosen to operate as a sole proprietorship, partnership, limited liability company, or corporation?
- Do you know the business and tax laws you will have to follow?
- Do you have a suitable location for your business?
- Are you prepared to maintain complete records of income, expenses, and accounts payable and receivable?
- Have you determined how much money you will need to start and maintain your business?
- Do you have the necessary person-to-person and telephone-sales skills to market your business?

If you are to become successful in your own business, you must set realistic short-term and long-term goals for yourself. The most effective way to do this is by writing a business plan. There are many books on this subject at libraries and bookstores, and a variety of business-plan software programs available on the market.

You can contact the Department of Economic Development in your state. Most have a package they can send you containing information important to small businesses. The Small Business Administration (SBA) in your area often offers free business consulting and accounting services to new businesses. SCORE, the Service Core of Retired Executives, can also provide similar services. Both organizations can help you write a business plan, provide a critique of the plan for you, and assist you with financial forecasting. The SBA can also provide you with potential loan sources.

Writing a business plan will allow you to forecast your earnings accurately over the next few years. It will demonstrate exactly what profit is needed to pay off your initial investment, buy additional systems, hire other entertainers, and expand your company.

Prior to starting your mobile DJ business, it is important to predetermine all of your pricing and services and to have prepared descriptions of them. In addition, it is wise to have thorough, legally binding client and employee contracts that have been approved by an attorney. It is very helpful to have a professionally produced promotional piece that is made available to clients on video or DVD.

Choosing a Business Name

Choosing a catchy and memorable name for your business is extremely important. Choose carefully because you will have to live with the name for a long time and build your reputation upon it. "Mike's Musical Madness" may appeal to you today, but is it the name you will want for your company 10 years from now? The name you select should be markedly different from those of any of your competitors and be as original as possible. Use it to reflect the "personality" of your company. It is not advisable to choose a name that is offensive to anyone or will appeal only to a small segment of the marketplace.

It is important to do a trade name search first and then to have a graphic artist create your logo. To obtain a trademark for a business logo, write to The United States Department of Commerce, Patent and Trademark Office, Washington, D.C. 20231. The current fee is about $350.00.

Registering Your DJ Service

If you have decided to work as a sole proprietor, you will want to ensure that no other company is using your chosen business name. After completing a name search at your county clerk's office, you can obtain a trade name certificate for a nominal fee. If you wish to incorporate your business for liability or tax purposes, I advise speaking to an attorney about what is involved in this process. You can incorporate without the assistance of an attorney or by using services that are available on the Internet.

Developing a Sound Financial Plan

Professional mobile disc jockeys generally invest significant capital into their companies. They are concerned with the continued growth of their businesses and with establishing a nest egg for retirement. Here are some suggestions to help you make the right decisions.

Consider your company as an investment vehicle. Seek guidance from a qualified and highly recommended financial planner. This person will help you determine the value of your assets and recommend suggestions for your diversified investment portfolio. A good plan keeps you from putting "all of your eggs into one basket" and helps you to achieve your financial goals over time. Good investment vehicles may include an IRA, mutual funds, stocks, bonds, or other investment vehicles. Keep a close eye on your portfolio to keep track of how your plan is going, and seek additional advice if the plan is not going well. Over time this financial planning process will enable you to rely less on earned income (that is, the income you derive from your DJ business) and more from unearned income.

When You Become a Golden Oldie

Future retirees, currently in their twenties, should consider starting to save now. If you are a part-time DJ and work for a corporation that offers a trusted 401(k) program, sign up. This will allow for a portion of your salary (up to 10 percent) to be deducted before taxes and invested into the program to accumulate tax deferred. Some companies will match all or a portion of your 401(k) contributions. You do not pay tax on this money until you withdraw it.

If you are self-employed, you should consider qualified plans such as simplified employee pension plans (SEP-IRAs), which offer attractive tax advantages and an individual retirement account (IRA). The earlier you start a retirement savings program, the better off you are likely to be in your golden years.

Accounting and Bookkeeping

When establishing your DJ business, be sure to seek guidance from a reputable accountant—preferably one who already has other mobile DJ clients. This person can advise you about legitimate business deductions, tax-saving opportunities, and can file your state and federal taxes.

If you have the skills to handle your own bookkeeping, there are several user-friendly bookkeeping software programs available on the market. If you cannot do the bookkeeping yourself, hire someone to do it for you. Some payroll services will also file quarterly taxes for businesses.

In addition to bookkeeping, payroll, and tax filing, you will need to establish a business checking account. It is from this account that all business-related bills will be paid and all business monies deposited. In the event that you reimburse yourself from this account for business expenses you have incurred, be sure to note in the memo portion of the check for what the check was written.

Examine your accounting when you receive your bank statement each month. Keep a detailed record of payroll and expenses. It is essential to keep accurate records of your profits and losses to keep abreast of the financial status of your business.

Opening a Merchant Account

In addition to setting up a business checking account, it is advisable to set up the ability to accept major credit cards as well. This will support your sales efforts because many clients will choose to pay for your services with a credit card. Offering this form of payment to your clients will definitely add to your bookings.

The first place to open a merchant account is at the financial institution where you normally do your banking. Past credit history is a major factor in determining how quickly you can open a merchant account. If you are not in a position to get an account immediately, find out exactly what you will need to do to get established so you can get this account.

You may need to make a formal appointment with a bank manager. It is recommended that you arrive in business attire with your business plan, marketing materials, tax records, and account information in a briefcase. Talk about your excellent account history with the bank and the substantial investment you are making in your DJ service. Explain how important being able to accept credit card payments is to your company's success. Present your business plan and projected profits.

If your banking institution turns you down, there are many other independent companies that offer businesses the ability to accept credit cards from their customers. Do your homework, and compare the percentage rates that these companies take from your profits.

Toll-Free Service

Establishing a toll-free service can provide a big boost for your sales if a large percentage of your clients must make long-distance calls to contact you. When establishing your basic and toll-free service numbers, create something catchy that incorporates your business name or the type of business.

Aside from the one-time installation fee and a flat monthly charge, the price for this service will depend upon where the call originates from and the day and time the call is made.

It is recommended to call several phone carriers to compare prices. If you conduct all of your business through one company, this carrier may be able to offer you a total business package at considerable savings. Be a smart shopper, and take the time to research the best rates and services for your company.

Legal and Accounting Advice

Smart businesspeople are aware of what they do not know and when to ask for help from an expert. The professional assistance you seek will depend on both your personal business skills and the availability of expert advice. Find professionals you like and trust in the planning phase of your mobile DJ service.

You can discuss the pros and cons of setting up a corporation versus a sole proprietorship, limited liability company, or partnership with an attorney. A lawyer can also look over your business contracts to make sure they are legally correct and binding in your state.

Profitable Additions to Your DJ Service

There are a number of products and services to consider adding to your basic mobile DJ service. These options enhance your ability to serve your customers and allow you to gain greater profits with your bookings.

Profitable additions to consider include:

- Karaoke
- High-energy pro dancers
- Props, prizes, and giveaways
- Big screen video/DVD
- Lighting and special effects
- Impersonators
- Magicians
- Clowns
- Comedians
- Hypnotists
- Games
- Carnival games
- Food concessions
- Casino equipment
- Sound reinforcement

Initially, you may want to subcontract services in which you do not have expertise or lease products that you do not own. Eventually you can bring these services in-house for even greater profits.

One way to make your service very attractive to a prospective client is to offer a complete party-planning service. You can be the client's "one-stop shop" for all of their party needs. To do this effectively you will need to establish business relationships with printers, limousine services, caterers, photographers, florists, and other providers of the services you wish to offer.

An attorney can advise you on the legal details of these affiliations, but generally you should receive a substantial discount from the proprietor for booking their product or service through your company. Try to arrange a deal in which the client pays you directly, and then you pay the proprietor within 30 to 60 days after receiving their bill.

Setting Up an Office

To keep your overhead low, you may want to consider initially working out of a home office and using a post office box as your business address or arranging rental space with a compatible business, such as a photography studio. You can even personalize your service by meeting with the client at his or her residence.

Here is a list of the business tools needed to properly function in a home office:

- Computer with word processing, accounting, scheduling, and travel directions software
- Laser or color printer
- Desk
- Bookcases
- Business phone and line
- Cellular phone
- Pager
- Fax machine
- Scanner
- Business checkbook
- Legal pads
- Paper clips
- Subscriptions to trade publications
- Filing cabinet
- Copy machine
- Answering machine or voice mail
- Mileage books
- Maps
- Marketing materials and company video/DVD
- Request and song lists
- Stapler and staples
- Scissors
- Pens and pencils

You can gain access to a wealth of important information by going on-line and using the Internet. The information on the Internet tends to be the most current and is quickly accessible. There are also DJ-specific "chat rooms" from which you can gain valuable insight and contacts. Currently, the most popular are www.themobiledjhandbook.com, www.djzone.com, and www.prodj.com. The World Wide Web is a great place to go shopping for new equipment and music.

If you run your business from your home, establish a separate business line with an answering machine. You may not appear professional to a potential client if a child, significant other, or any other individual not involved with your business answers the telephone. Some callers will have entertainment emergencies and will need to get in touch with you immediately. You may want to include a pager number on your answering machine message to address this situation. Your answering machine message should be brief and delivered in an upbeat, professional style. Here is an example:

Thank you for calling The Best DJ Service, the largest mobile DJ service in Kansas. This is Bob Simmons, and although I'm not available at the moment, your call is very important to me. So please leave your name, telephone number, and any message after the tone, and your call will be promptly returned. If this is an entertainment emergency, you can page me at (number). I'll get back to you immediately. Thank you for calling The Best DJ Service.

This message does not intimidate the caller. The idea is to get potential clients to leave their name and telephone number. Make it as easy as possible for them. Some DJ companies prefer using an answering service rather than an answering machine. It is sometimes difficult to find a quality service that will handle your company's calls in a manner that is prompt, courteous, and professional.

Tracking Clients and Gigs

There are software programs available that are designed specifically for mobile DJs. Most feature a screen with all of the fields necessary to take client information over the phone. Some of these fields can be customized to suit your own needs. They also include the abilities to apply the information from your initial data entry and to print a contract. The contract form can be customized to include your terms and conditions.

Using a DJ-specific database program reduces duplication of effort and creates a "paperless" office environment. You need only input the client's name and address once, and from that information, print out an envelope, contract, cover letter, invoice, or a variety of other forms. Additional information such as payment arrangements and a client's music preferences can also be entered.

A calendar function allows you to look up a certain date to review your bookings and the client's status, such as "pending" or "confirmed." A venue database section can contain directions to each of your regular event facilities. There is also a feature that allows you to view all pertinent client information on one screen.

Letterhead and Contracts

Your DJ company will need letterhead and envelopes that coordinate with your business cards. It is important to maintain a consistency of quality and reflect your company's image and professionalism. Use a high-quality bond for your letterhead and envelopes. Ask at least three printers to provide you with quotes for doing your business cards, letterhead, envelopes, and marketing materials.

Obtaining Insurance

Insurance is a necessary expense to ensure that you are legally and financially covered for every contingency related to your DJ business. The kinds of insurance you, your employees, and sub-contractors may need include worker's compensation, business liability coverage, auto liability, equipment/contents coverage, loss of income/business interruption, life insurance, health insurance, and disability. Make sure your equipment is fully covered, regardless of its location. Secure the type of insurance that provides for current replacement costs.

If possible, find out the names of the insurance companies that your competitors are using. This is important because these companies are familiar with the mobile disc jockey business and will properly insure you. A good resource for obtaining insurance company information is through national and local DJ associations.

For your insurance files, make a visual record of all of your equipment and music with close-ups of the serial numbers. This can be very helpful in the event your property is ever lost or stolen.

For insurance purposes, it is important to know that the U.S. Department of Labor recognizes our profession as a legitimate vocation. The occupational title is Disc Jockey (Mobile), Occupational Code # 159.142-560.

Buying a Franchise

For a franchise to be valuable, the franchise must have name recognition or a proven business system. While some people enjoy the challenge of starting a business from its infancy, others are more comfortable purchasing a proven success formula.

As a franchise owner, you will immediately take a managerial role in an existing, established business. You will be paying a fee, so there should be a guarantee that the franchise will save you money in some areas and keep you from making costly mistakes in others. Before making a final decision about buying a franchise, thoroughly examine the operation. Make your own determination about whether the knowledge, experience, mode of operation, and reputation makes the purchase worthwhile. Insist on documentation that substantiates any claims that are made. Go over all of your findings with an attorney before signing on the dotted line.

It's nice to be important,
but it's more important to
be nice.

JOHN CASSIS

2

Getting Top Billing

In the DJ business, perseverance is always the key to success. However, you also need to have a passion for music, the talent to play it well, and the ability to stay open minded. Study, absorb, and work hard until the results come. Keep doing this again and again. Before you know it, people will begin to believe in you more than you could have expected in your wildest dreams!

FRANK GARCIA, OWNER
Mainline Pro Lighting & Sound
New York DJ Entertainment School

Creating a Reputation

The best mobile disc jockeys constantly strive to improve their performance. They take pride in their work and are committed to their clients' satisfaction. They endeavor to achieve a level of excellence better than the best; they understand that success is based not only on talent, but also on persistence, drive, and business acumen.

Working long hours on your feet while keeping a high-energy level is not easy. Achieving and maintaining a positive attitude can be a challenge. But each time you put on an outstanding performance it will be noticed, and folks are likely to spread the word about the great job you did.

Energize your mind with books, CDs, tapes, and activities that will motivate, relax, and inspire you. Eating healthfully and getting regular exercise will also help keep you in peak, physical condition.

The greater your number of bookings, the more exposure you will have. The more exceptional your performances, the more likely it will be that people will pick up your business cards and call you or recommend you to others.

As a DJ entertainer, it is vitally important for you to be in touch with your audience's desires and to play solely for their enjoyment. Play requests within a reasonable timeframe. If the guests at an event are not dancing to the music, do not look happy, or are complaining to you, CHANGE WHAT YOU ARE DOING IMMEDIATELY! A happy client or guest may hire you again. Likewise, an unhappy one will not.

Prior to an event, it is an excellent idea to clarify exactly what level of energy and participation your client is expecting from you. The answer you receive will give you valuable insight as to what your client has in mind for his event. Be flexible with your performance style. Always strive not only to meet, but to exceed your client's expectations.

Preparing for Events

Always arrive at any occasion with ample time to set up and test your equipment and to use the restroom to check your appearance. Do not consume alcohol before or during a job. This can negatively affect your performance and make a bad impression on your client and the guests. It is a good idea not to smoke in the audience's view.

It is essential to prepare for all engagement details prior to your performance. Once a signed contract and deposit are received, your pre-event preparation will begin. Develop a pre-event checklist that includes the following:

- Reserve the equipment and personnel necessary for the event. If a subcontractor is needed or outside equipment rentals are required, reserve these as well.
- Reserve the transportation vehicle for the event.
- Obtain directions to the facility, including the best equipment access route to the room where the event is being held.
- Contact the facility manager/representative, and discuss the details of the event, including the earliest access time you will have to the facility. Confirm your electrical and space requirements.

- Make any purchases necessary to fulfill contractual obligations with your customer.
- Ensure that the customer has returned a completed list of requested music by the week preceding the event.
- For an event that requires the DJ to act as an emcee, create an itinerary outlining key moments and the order in which they will occur. Confirm the itinerary with the customer and the facility contact person.
- Send job sheets containing pertinent information to all personnel participating in the event.

Further preparation occurs at a facility. A complete on-site checklist includes the following:

- Set up all equipment prior to arrival of the client and the guests.
- Test all equipment.
- Tape down all cables and cords.
- The DJ should arrive approximately one hour before the scheduled start time and immediately introduce him or herself to the client. Thoroughly review the event itinerary with the facility contact person, photographer, and/or videographer.
- The DJ should always dress in appropriate attire. Before your performance begins, be sure to set up the company banner and place small stacks of business cards on the speakers and/or tables.
- The DJ should stand and play continuous music during the event.
- Near the start of an event the DJ should introduce him/herself. At this time the facility name and the names of the photographer and/or videographer should be mentioned. Near the event's conclusion, this information should be repeated. When appropriate, the host or hostess of the event and the facility staff may be thanked.
- Always be acutely aware of your volume level. This is the most common complaint among clients and guests. Your treble and bass should be carefully monitored as well. If one person makes an isolated complaint, but it appears that your client and the rest of the guests are happy, it is probably not necessary to make any adjustments.

Figure 2.1 Christy Lane and Dancers. Courtesy *Mobile Beat* magazine.

Professional Attire and Etiquette

As a professional mobile DJ you must always be aware of how you appear to your audience. Failure to pay attention to your appearance may detract from your image as a professional and may affect future referrals.

Appropriate attire is essential for all bookings. It is necessary to invest in a black tuxedo jacket with matching pants. You will need a couple of white tuxedo shirts and a black bow tie, cummerbund, and formal leather shoes to go with these. Other nice additions to your attire can include formal and snazzy vests, and a red bow tie and cummerbund. For less formal occasions, a black or navy blazer with a white shirt and gray or khaki pants may be considered. Female DJs can design more feminine versions of these outfits, if desired. Find out what attire is expected by the client for their event.

Dealing with Clients

From time to time you will have to deal with difficult people. Some may be critical, demanding, or obnoxious. These times will test your patience. It is important to wear a smile and maintain a thoroughly professional attitude. Be as cordial as possible, but if things get out of control, you may want to elicit help from your client to handle the situation.

Customer Service

Rule number one for providing excellent customer service is to give the customers more than they expect. Go ahead, promise the moon and the stars—but only if you can truly provide them. If you give only lip service and do not follow through on your commitments, then you are still providing customer service—the bad kind.

Giving customers more than they anticipate will not only gain you repeat bookings but it will also enhance your reputation and lead to referrals. When clients are impressed, they tell their friends, colleagues, and acquaintances. When they are *not* impressed, they tell even more people.

When a client hires your DJ service, they automatically expect reasonable rates, a complete music selection, a professional performance, and pleasant service. Go the extra mile. Promise less and deliver more, then watch the customers line up at your door.

DJ Associations

An ever-increasing number of professional mobile disc jockeys are joining both local and national DJ associations. This is an excellent trend because it provides opportunities for us to pool our ideas and efforts to improve the level of professionalism within our industry. The standards bar keeps getting raised, and that is good for our profession. When thousands of association members make their collective opinions known about important industry issues, they can have a loud and powerful voice.

There are many benefits to being part of a pro DJ organization. The four current American national associations are:

- American Disc Jockey Association
- National Association of Mobile Entertainers
- Online Disc Jockey Association
- United States Mobile Entertainers Association

In Canada the associations are:

- Canadian Disc Jockey Association
- Alberta Association of Mobile Entertainers

Many individual states also have their own organizations. Some national associations have local subdivisions, where members can get together and work to benefit local professional causes and issues. One of the most important benefits in joining a DJ association is having access to equipment in the event of an emergency. Also, association Web sites, which inform the consumer, often help members generate extra referrals because an e-mail or a telephone number can often link a potential customer to your business.

Local association meetings typically include technical clinics, new dance-step training, and discussions about ways to raise professionalism and to combat bottom-feeders in the market. There is a friendly sharing of information, camaraderie, and even field trips. Some associations also participate at local charity events.

DJ Conventions

Attending industry conventions can change your professional life for the better and provide you a gigantic international networking opportunity. Turning out for these fun and action-packed events is essential for DJs who want to keep on top of the rapid changes that continually shape the scope of our business.

A visit to the convention exhibit halls offers mobiles the unique opportunity to see, hear, experience, and discuss the offerings of hundreds of vendors, all in one place. Trade shows also provide an opportunity for you to tell vendors and the show's organizers about your needs, opinions, and experiences. Your input can result in products and services that better meet the needs of our profession.

Training seminars and workshops offer wonderful learning opportunities. Vendor-sponsored parties provide a refreshing social outlet free of selling pressures. Special parties and social events are a great way to meet and speak with peers under casual, enjoyable conditions.

Figure 2.2 Author, Stacy Zemon, autographing a copy of *The Mobile DJ Handbook*, first edition, at the 1998 International DJ Expo, Trump Taj Mahal, Atlantic City, New Jersey.

Upon returning home from a DJ convention, be sure to organize the business cards you have received as well as the information you have taken from vendors. Review any notes you have taken and create an action plan from what you have learned. Make the most of your trade-show experience.

If you never have, you should. These things are fun and fun is good.

DR. SEUSS

3

The Interactive DJ

To be a great interactive DJ you must have the ability to "break the ice" with a group of total strangers. When you love what you do and are having fun, this will come across in your words and actions. Personalize the experience for your client. Demonstrate your best skills at being a high-energy party host who excels at teaching dances, contests, or games.

LISA "DO THE DANCE" CAPITANELLI
President
I'm A Girl DJ Entertainment
Los Angeles, California

Getting the Crowd Going

Sometimes people need to be coaxed to get them up and dancing. Some folks are shy about being the first on the dance floor. This is where your skills as a motivating party host are needed. Participation dances almost always work as icebreakers for difficult crowds. Find creative ways to get people to party!

One great technique to get the crowd going is to call up the host or hostess and announce that the conga line will form behind him or her. Another idea is to call upon your host or hostess to come up to the dance floor and to bring four people along. Send those four people out into the crowd to get four more. Continue this until your dance floor is filled. Remember—people *want* to have fun! You must relax *and* motivate them, so they can enjoy themselves. You can begin an event with an upbeat song that will appeal to the

Figure 3.1 The crowd tries to follow the smooth moves of Lisa "Do the Dance" Capitanelli, president, I'm A Girl DJ Entertainment, Los Angeles, California.

majority of the audience. This will help set the tone for the rest of the engagement.

Demographics such as age, race, economic class, ethnic background, gender, sexual orientation, and geographic area are factors that determine the types of music people like. One of the purposes of a DJ apprenticeship is to provide you with the exposure to a variety of audiences. This experience will help you determine what types of music various groups will appreciate and will teach you how to "read a crowd."

While performing, mingle with the crowd during the cocktail and dinner hours. This socializing may help the guests warm up more readily. Talk with people and ask for their feedback. Write down their requests or give them a tabletop request card. When the dancing starts, let the audience know that you take requests.

The music played at an event can cover a range that spans almost every conceivable style and decade. The key to success is to do your best to please all segments of the crowd. If the majority of an audience does not enjoy a particular musical style, stay away from it. A tactful way to tell someone "no," is to say that you did not bring the song with you. Only play requests for songs or types of music to which you determine the majority of guests will want to dance.

Microphone Skills

Your ability to communicate and motivate over the microphone is second in importance only to your music programming abilities. Acquiring this skill requires significant practice, and there are several ways to continuously improve your presentation abilities.

Observe other DJs in action. Take notes about what they say as well as how and when they speak. Practice at home speaking into a tape recorder. Apprentice at events and get on the "mic" under the watchful eye of a video camera. Review this tape later, paying close attention to your facial expressions, speech mannerisms, inflection, accent, and vocal qualities. Some good questions to ask yourself (and your peers) are: Do I project well? Can I be easily understood? Do I sound energetic and enthusiastic? Am I speaking slowly enough? Do I look happy?

To be clearly heard when speaking, it is important to keep your mouth close to the microphone and to have your mixer turned up to an adequate volume level. Generally, begin all of your announcements with "Ladies and gentlemen, may I have your attention please?" If people do not stop talking, say this again with more authority. Be sure you have everyone's attention before getting to the heart of your announcement.

Remember to promote your DJ service at the start and end of an event. Mentioning it one or two times during the party is fine as well. The best time to do this is at the end of a successful dance set. This way, potential new clients will equate you with fun.

Write down and spell names phonetically before announcing them to ensure that you pronounce each one correctly. Plan out in advance what you are going to say. For beginners it is best to write *everything* down, prior to turning on the mic. Do not try to wing it until you have polished your microphone presentation skills to the point where you shine like a diamond.

Ice Breakers

Many people are shy and awkward about being the first ones on the dance floor. The secret to your success at any affair is to break the ice early. You will have to dissipate the invisible barriers that commonly exist in these situations and be an interactive party host/crowd motivator who personifies the word "fun."

Dances, Contests, and Games

One of the most effective ways to please a crowd is by incorporating participation dances, contests, and games into an event. These activities offer people an opportunity to mingle and, for some, to show off. As the DJ and emcee of an event, you will announce, explain, coordinate, and demonstrate how activities are conducted. There are game books on the market that can give you some great ideas. The archives of trade publications and DJ Internet sites such as www.djzone.net also have excellent listings of new and innovative ideas.

With some contests, you can be the judge. In other cases, you can select judges from the crowd. There will be occasions when you may need only announce an activity once and people will immediately participate. At other times this will not be the case. When this happens you must show your enthusiasm, step out into the crowd, call upon your client to assist you, and be creative to get people involved.

Participation Dances

Leading a couple of participation dances at an event is almost always a sure-fire icebreaker and crowd pleaser. Know how to teach the classics and be sure to stay on top of what is new. You can even invent your own dance if you are so inclined. There are videos on the market that provide step-by-step instructions for learning participation dances. New dances are unveiled every year at DJ conventions.

When providing instruction for participation dances, pay close attention to how much instruction is needed. If roughly 75 percent of the crowd is correctly matching the steps you are teaching, then it is time to let them dance along to the music.

Anniversary/Marriage Countdown Dance (Popular ballad)

This dance is primarily used at wedding receptions. Invite all of the married couples and the newlyweds to the dance floor. Request that the newlyweds remain throughout the entire dance. Every thirty seconds to one minute speak over the music and ask everyone that has been married a year or less to please leave the dance floor. Keep counting by fives until you reach 40 years. Then count "41, 42, 43. . ." until only the longest-married couple and the newlyweds remain. You might walk out to the couples and ask the crowd to applaud them. Ask the older couple to share their secret for a successful marriage. Hand them the microphone so the whole party can hear their answer.

Multiplication Dance (Popular upbeat song)

Everyone stands in a circle for this dance. Choose a man and a woman to dance in the middle of the circle. Announce that when you say "break," the couple dancing in the middle of the circle should then each pick someone of the opposite sex to dance with. Only the people in the middle of the circle dance. This process continues every time you say "break" until everyone is dancing.

Line Dance (Hang On Sloopy, The Stroll)

Ask participants to form two lines facing each other. All of the guys stand in one line and all of the gals stand in the other. Instruct the couple at the far end to dance between the lines, bumping their hips along the way, when the music starts.

Dollar Dance (Popular ballad)

The Dollar Dance is traditionally done at a wedding reception. For a dollar or more, any guest can briefly dance with the bride or groom. To begin, ask the maid/matron of honor and the best man to assist you by standing next to the bride and groom, respectively. Ask the women participating to line up behind the groom and the men, behind the bride. Play a popular ballad or two to give all the participants the opportunity to dance. It is the job of the best man and the maid/matron of honor to collect the money and to be sure that no one dances with the bride and groom for longer than 30 to 60 seconds.

Snowball Dance (Popular ballad)

Select a man and a woman to start the dance. If there are guests of honor, choose them. Play a song for approximately 30 seconds,

Figure 3.2 Do a little dance, make a little love, get down tonight! Oh, how DJs love to learn new dance steps at the *Mobile Beat* DJ Show and Conference. Courtesy *Mobile Beat* magazine.

then call out "Snowball." The couple then selects another couple and brings them onto the dance floor. This process is repeated until the dance floor is filled.

Contests and Games

The following activities can be incorporated into any appropriate event and often are used at Mitzvahs.

Best dancers contest

Pick three judges. Explain to the crowd that the contest is both for couples and for singles, and that the judges will choose the best 12 dancers by tapping individuals or couples on the shoulder during the first half of the song. Fade down the music after the song is half over, and ask everyone who was not tapped on the shoulder to form a circle around the dance floor. Continue to play the second half of the song. The judges will choose the best three couples or singles remaining and tap them on the shoulder. When the song ends, those who were not tapped on the shoulder will join the circle. Those standing in the circle will applaud for each couple or dancer. The volume of the applause will determine the third-, second-, and first-place winners.

Hula hoops (*The Twist, Twist and Shout, Let's Twist Again*)

Divide the group into teams of 5 to 10 people. Invite one team to start the contest by forming a line. Give each member of the team a hula hoop. Explain that when the music starts, they must spin the hoop continuously between their shoulders and knees only. A contestant is out when his or her hoop falls to the ground. When there is only one person left, that team is finished. Repeat the exercise with each of the teams. After all of the teams have played, instruct the winners of each round to form a large circle for the "Hoop Off." If you want to add a degree of difficulty, you can ask the contestants to spin the hoops while standing on one foot.

Freeze contest (Popular, upbeat song)

Instruct people to stand in the middle of the dance floor and to dance in couples or singles while the music is playing. When the music stops, they must immediately freeze, or a judge will tap them on the shoulder, and they will be "out." Award prizes to the third-, second-, and first-place winners.

Trivia night

Either teams or individuals can enter, but you can only have one team per table. Hand out paper and pencils, and ask a few trivia questions at the start of each song. Entrants have until the end of the song to write down their answers and bring them to you. Use various topics, such as sports, entertainment, music, and so on. Assign a number of points for each correct answer. For the final question, have the teams wager points on their final answer.

Huggy Bear (Popular, upbeat song)

Call participants to the dance floor. Explain that when the music starts everyone must begin dancing, and that you will be stopping the music periodically and calling out a number. Everyone has to form groups of that number as quickly as possible. Anyone unable to get in a group for that number is out. This continues until there are only two people left. They are the winners.

Back-to-back (Popular, upbeat song)

Ask people to dance as singles on the dance floor. When the music stops, everyone finds a partner and interlocks their arms back-to-back. The last two people to form a pair are out. Start and stop the music until you have one remaining pair on the dance floor. They are the winners.

Limbo (*Limbo Rock*, Caribbean music)

Have two volunteers hold a six-foot pole at each end. Distribute leis and ask people to line up in front of the pole. Explain that when the music starts everyone must dance under the pole by arching their backs with their heads being the last thing to go under. Contestants who bend forward or arch their necks will be "out." Leave enough space between the participants so that they do not bump into each other. Every minute or so, lower the pole a few inches. Eliminate people when they bend improperly, touch the pole, or fall.

Single mingle night

This is primarily done at pubs, clubs, taverns, and bars. Distribute two different-colored prize cards to each person. The men get one color, the women the other. Each card has two prizes written on it, such as a pair of concert tickets, a TV, a $25 bar tab, or a vacation. As people meet, they will compare their cards to each other. If they match prizes, they both win.

Musical chairs (Popular, medium-beat song)

This game is primarily conducted with youngsters. Line up chairs back-to-back and side by side in a line. There should be the same number of chairs on each side, and the number of chairs should equal the people participating, minus one. Instruct people that when the music starts they should walk quickly around the chairs in a clockwise direction with their hands on their heads. Explain that when the music stops, they should take a seat, and if they are left without one, they are out. Start the music and repeat the process removing one chair at a time until only one person is left.

Coke and Pepsi (Popular, high-energy song)

This is a relay game that is great to use with youngsters. Form two lines with an equal number of people. Ask them to stand across the

Figure 3.3 DJs like to party too! Kickin' it up at the *Mobile Beat* DJ Show and Conference. Courtesy *Mobile Beat* magazine.

floor facing each other. Designate one side as Coke and the other as Pepsi. Explain that when you say "Coke," the Pepsi side must run and sit on the knees of their counterpart on the Coke side. The Coke side must squat down so they can do this. If you say "Pepsi," the Coke side must run and sit on the knees of their counterpart on the Pepsi side. The last person to be seated is eliminated from the game. If you say "Dr. Pepper," the groups switch sides. If you say "Hi-C," participants give their counterparts a "high five" and return to the side of the room they were on. If you say "Wayne's World," everyone bows, placing their hands over their heads then bring them down to the floor saying, "We're not worthy." If you say "freeze," everyone must stop dead in their tracks. If you say the name of the guest of honor, everyone must point to that person and say, "You're the best." Everyone should dance in place between orders. After each instruction, those who do not comply are eliminated. The game is continued until there is a winner. You can add your own variations to the game, such as using inflatable plastic feet as props.

Simon says

Children or adults stand in a line on the dance floor, facing the DJ. They respond to your commands such as, "Simon says left hand out" or "Simon says spin in a circle." This process continues and people are eliminated from the game if they follow a command that is not preceded by "Simon says." The contest winner is the last person remaining.

Mummy wrap

Pairs of children are each given a roll of toilet paper. One child is the "mummy." The object of the game is for the other child to wrap up the mummy as fast as possible without tearing the paper. If the paper breaks, that team is out. The first team to successfully use the whole roll without a tear is the winner. If no team can do this, then the last remaining team is the winner.

Scavenger hunt

Giveaway items such as toys or candy are hidden around a specific area in an event venue, in difficult but not-impossible-to-find locations. Children are given buckets to use and whoever gathers the most items in a preset amount of time wins the hunt.

Balloon T-shirt

This game requires that you supply XXX-large T-shirts for the number of people playing the game and a significant number of 9-inch balloons. The children form small groups to play against each other. Typically, the smallest child is the balloon T-shirt person. Tell the children that the game's object is to blow up as many balloons as possible and then stuff them under the T-shirts while you play a 3 to 4 minute song. The group with the most balloons stuffed in the T-shirt at the end of the song is the winner. Calculate this number by counting the balloons as you pull them out then pop them with a pin. Do not pop the balloons while they are still under the T-shirt because this may sting the child.

Truly interactive mobile disc jockeys always have the opportunity to be creative by inventing their own new games and contests, or by taking a traditional game and adding a unique twist.

Karaoke

KJs use many of the same equipment components as DJs. To offer Karaoke as an entertainment option, you will need a CD+G player,

which processes the graphics for the songs on the CD+G discs
you will also need to purchase. Karaoke DVD sets are currently
emerging into the marketplace. You will also need to provide tele-
visions if the venue where you are performing does not have them.
You will probably want to have a stable of at least three microphones and several up-to-date songbooks. Be sure to have pencils
for people to write down their requests. Whenever possible, create
your set up in a manner that limits visual distractions behind the
performers.

To be a truly professional Karaoke jock, you must conquer the
invisible boundary with your audience and truly believe that there
is nothing more enjoyable than watching and listening to people,
who have had too much to drink, sing. This "enjoyment" usually
includes inaudible and poor performances, the dropping of and
screaming into your expensive microphone, and a wide array of
spontaneous activities that will keep you on your toes. Don't lose
your temper—no matter what happens. Remember, you chose this
business because it is fun, right?

Depending on your crowd, you may get a lot of singers, or
you may get a lot of people that want to drink and dance. Even shy
people may be willing to sing if given some assistance and coaxing.
Always bring the eager volunteers onto the stage first. If the "well"
dries up, try wandering the floor with your wireless mic. As you
sing, notice the people who are also singing at their tables. Stop
there and hold the microphone halfway between their mouth and
yours. This puts people at ease. Make the experience fun with no
pressure.

Use your own creativity to build on and create your own
"schtick." Be sure to intermingle dance sets with Karaoke sets.
Always let your singers know who is coming up in the next set.
For some mobiles, Karaoke can be an excellent adjunct to your DJ
business.

Party Props

One of the most important tools in interactive DJ-ing is the use of
props. These are the fun items people use as well as people themselves. Get your audience involved in the action whenever possible.
This can include getting them to lip-sync, wear costumes and wigs,
and jam with party props such as inflatable instruments.

Design an interactive program that incorporates appropriate props for the type of engagement at which you are performing. The primary factors to consider are demographics, logistics, theme, and budget.

There is a universe of party prop choices and companies to buy them from. Themes can include YMCA, Motown, surfing music, British invasion, and more. The most popular props include items such as costumes, wigs, hats, big foam hands, rope lights, inflatable and plastic instruments, sunglasses, limbo pole, leis, hula hoops, glow items, conga sleeves, masks, bead necklaces, hula skirts, and kazoos. What's more, there are also games such as *Twister*, *Jeopardy*, and *Trivial Pursuit* that DJs commonly use with party guests. Creative Imagineering offers "Game Show Mania" products that allow you to turn an ordinary function into a television trivia game show event. While these products can be costly, they offer great upsell opportunity to your DJ service.

Party props in general offer increased revenue potentials for your business. Purchasing them by the case will yield substantial savings. Your cost versus your sales price makes this volume purchase worthwhile. If you plan to use a large number of party props at an event, it is a good idea to organize them before your performance starts. Large plastic bins or crates work well for storage. Clearly note on a label the contents inside.

Bar/Bat Mitzvahs generally incorporate the use of more party props than does any other type of event. Professional high-energy dancers at Mitzvahs are very helpful for distributing party props quickly to guests. You or the dancers need to make clear which are personal-use props and which are guest-use props. The first category needs to be returned and the second category includes items that are available to take home.

If your mixer has a digital sampler, you can use this as an interactive tool. Just record some famous spoken phrases, or a few bars from a TV theme or popular song, then play them back to the crowd at high or low pitch, or fast or slow speed. Ask people to guess what the phrases are. You can even hand out prizes to those who guess correctly. People love to win prizes, and they need not be expensive.

If I were a toy, I would say,
"Press my play button."

JULIE DeKOVEN

4

Getting Geared Up

*Everything breaks! Especially being constantly moved
around from venue to venue and event to event. You always
need to carry spare gear with you so the show can go on. Choose
your equipment with the same considerations that mechanics
choose their tools focusing on quality, reliability, and
comfortability. Remember, technology is great—when
it works!*

BERNIE HOWARD, OWNER
Bernie Howard Entertainment, Inc.
Northbrook, Illinois

Where to Buy

Your single largest expense in starting a mobile disc jockey service
will be for the purchase of equipment. Thorough planning and consideration should be given to what you will need. Determine what
your cost will be, and then find a way to afford it or wait until you
can. Base your equipment decisions on the requirements of events
that you plan to book. If you want to specialize as a high school
event DJ, you will require more gear than if you decide to focus on
200-person weddings.

There are a few different ways to purchase equipment. You can
work out a package deal with a DJ, pro-audio, or music store that sell
disc jockey equipment. Stores that specialize in DJ gear are located
in most major cities. These businesses often have salespeople who
are disc jockeys and who are willing to lead you toward the best
equipment values. These businesses will usually offer a discount if
you buy all of your gear from them. The primary benefit to buying

Figure 4.1 Bernie Howard, Owner, Bernie Howard Entertainment, Inc., Northbrook, Illinois.

this way is that these stores allow an exchange period. This is a plus if you are not delighted with your purchases. These stores usually can supply you with a loaner if a piece of your gear fails (and you have an engagement booked) and also provide on-site repair service.

Another option is to buy through mail order or via the Internet. There are several discount sound and lighting equipment companies that cater to DJs. When working with these dealers, you need to know exactly what equipment models you want. You usually cannot return a product because it is not what you thought it should be. These companies sell new equipment in sealed boxes, and if you have a problem with the product, they urge you to send it directly to the manufacturer for repair. They advertise in trade publications and in DJ and music magazines. These companies usually offer significantly lower prices than when buying retail. Sometimes, as long as the purchase is shipped from another state, you do not have to pay tax on items you buy through mail order. This can add up to a lot of money saved when you are spending several thousand dollars; however, shipping charges usually apply.

Be careful of shipping and handling charges. Negotiate the full amount of shipping—including insurance—before giving anyone your credit card number! The downside to buying mail order is that these companies usually do not offer any kind of service agreement; the equipment carries only the manufacturer's limited warranties. Compare carefully before you buy.

If you cannot afford new equipment and need an item immediately, there is good used·equipment available on Ebay (www.ebay.com), in local newspapers, trade publications, and bargain-hunter guides. When buying used equipment locally, set it up and test it for performance. Be sure the equipment comes with an owner's manual and warranty. The price you pay for used equipment must be far less than original retail to make it a worthwhile purchase. If the seller has the receipt for his original purchase, the warranty may be transferable to you. Another option is to rent equipment from a DJ supply store for each job until you can afford to buy your own.

Expand your knowledge of the equipment available in the marketplace by gathering DJ catalogs containing professional sound and lighting equipment. Be sure the amplifier and speakers you buy are compatible with one another. You can save money by buying a mixer without sampling capability. Perhaps you will want to use the savings from this to buy another needed component.

Warranty cards are important to manufacturers because they want to know why you bought their gear. Send the cards in, but more importantly, retain your receipt of purchase. Should your equipment require warranty repair, the receipt will show the date and the dealer from whom you bought it. The sales receipt should also contain the serial number (for insurance and warranty purposes). If your equipment ever needs to be repaired, photocopy the receipt and send the copy along with the unit to the manufacturer. Never send your original receipt. It is important to retain all original receipts for tax purposes, insurance, and warranty repair.

Sound Equipment and Accessories

A good sound system is reliable, durable, compact, and easy to operate. Consider purchasing the following components and accessories:

- Two-way or three-way speakers with 15-inch woofers that can handle at least 250 watts of power. (Speakers are actually pistons that move air—bigger is usually better.)
- Amplifier with power of 50- to 100-watts greater than your speakers are rated (Having amplifier headroom is a good thing.)
- Speaker stands with bags (for speakers under 100 lb)
- 1 power conditioner and light module
- 1 mixer
- 1 sonic maximizer, equalizer, or aural exciter
- 1 dual transport compact disc player
- 1 compact disc player and cassette deck combo unit
- 1 cassette player/recorder
- 1 MD recorder/player
- 2 turntables
- 1 wireless microphone and receiver unit (UHF frequencies are less populated than VHF; VHF is always cheaper)
- 1 mixer light
- 1 wired microphone
- 1 microphone windscreen
- Microphone stand with boom
- 1 single muff or dual muff cueing headphones
- 1 rack
- Cases for all components
- 1 compact disc case with divider cards (A–Z, "Soundtracks," "Ethnic," "Gotta Play," etc.)
- Viewpaks for your CDs (These are much thinner than jewel cases.)
- 1 compact disc player lens cleaner
- 1 Swiss army-type knife
- 1 3-prong AC adapter (floating the ground can get rid of hum)
- 1 cassette demagnetizer

Accessories

- 1 equipment cart
- 1 accessory/tool case
- 9-volt alkaline batteries or AA for wireless mics
- 1 large plastic tarp

- 1 50-ft black extension cord
- 1 25-ft black extension cord
- 2 cube tap outlet splitters
- 1 wire stripper and clipper
- Several rolls of duct tape and electrical tape
- 1 Phillips, flathead, and banana plug screwdriver
- 1 pliers
- 1 scissors
- 1 volt/ohm meter
- 1 polarity tester
- All necessary cables, conductors, and plugs
- Spare fuses

You may also want to consider adding a complete Karaoke system to generate an additional revenue source for your business.

Lighting Equipment

While standard DJ audio equipment can run through one 15-ampere circuit, lighting requirements are much different. Up to four or more separate circuits may be needed to run a mobile light show.

Most commercial buildings (schools, hotels, banquet halls) use 20-amp circuit breakers to the wall outlets. This means that the wire in the walls will handle 20 amps of AC draw. Residential codes only require wire to carry 15 amperes. To operate safely you will need a separate circuit for every 1000 watts of lighting. When you book a light show you must find out if the venue can accommodate your electrical needs. When testing your gear before an event, plug in all of your lights and play music loudly to load the circuits. This is how you will discover if the breakers will hold. To avoid hum in your system from lighting dimmer noises, do not plug lights into the same circuits as audio.

Lighting can be an additional profit center for your mobile DJ service. The ability to offer a customer a lighting package can sometimes help close a sale. If you offer low-end, mid-range, and high-end packages, you will be well prepared for any type of event. Lighting takes more time to set up than audio does, so don't forget this when planning your set-up schedule.

Lighting is classified by two primary categories: intelligent and nonintelligent. Nonintelligent lighting is the lesser expensive and easier to use of the two. There are no special cables or programming involved. This equipment is also lighter in weight, yet all lighting is very fragile. It is wise to invest in lighting cases; the protection they provide is worth the cost.

Nonintelligent sound-activated lighting has a small microphone built into the unit that "listens" to the beats in your music. The sound triggers the motor that moves back and forth to the beat.

Intelligent lighting is programmed by the user and is controlled by a computer. This allows several units to be used simultaneously while having one central controller. This type of lighting is used to create or enhance a mood through preprogrammed patterns. It is the most expensive type of lighting and is most often seen in clubs. Mobiles can use intelligent lighting to highlight special points in the room (i.e., the wedding cake or the doorway where the bride and groom will enter). Preprogramming these locations can allow you to make introductions much more dramatic.

There are many manufacturers, retailers, and types of lighting to choose from in the marketplace. It is recommended that you visit major lighting dealers to see these special effects demonstrated, or to view videos from manufacturers. It is difficult to select effect lighting from only a picture on the Internet or in a catalog. It is important to have a complete understanding of how to set up your lighting system, the voltage and amps required, and all the relevant safety information. You may want to put the system together yourself while an expert guides and watches you. Repeat back to him or her, step by step, what you have just learned. This will ensure that you have a complete understanding of how the equipment works and that you will be able to instruct other members of your company. You may want to videotape this process for future reference.

Here are some lighting and effects items to consider:

- 1 12-inch mirror ball with stand and rotator
- 2 pin spots
- 1 12-inch police beacon
- 1 strobe light with variable speed control
- 1 light stand
- 1 lighting truss
- 1 fog machine
- Rope lighting
- 1 bubble machine
- 3 sound-activated lighting effects

Party Favors

People can act zany when you give them party favors. Whether at a wedding, corporate party, or Mitzvah, party props help people get into the mood to celebrate. You can sell them to your customers, or suggest a purchase to the party host/hostess.

Some examples are:

- Hawaiian leis
- Sunglasses
- Balloons
- Inflatable guitars and saxophones
- Bubbles
- Hats
- Maracas
- Rubber masks
- Various neon items
- Limbo pole
- Hula hoops

Personal and Travel Items

- Tuxedo or sport coat
- Dress shoes
- Cummerbund, vest, bow tie, and shirt studs
- Deodorant
- Cologne or perfume
- Brush
- First aid cream
- Bandages
- Hairspray
- Breath mints or spray
- Pain reliever
- Gloves and scarf
- Money
- Credit card
- Sewing kit
- Mace or pepper spray
- Umbrella
- Hat
- Sunblock

- Insect repellant
- Automobile club card
- Gas credit card
- Jack and handle
- Inflated spare tire

System Components

The following are types of gear that need to be considered in creating a mobile DJ rig.

Mixer

A mixer controls all of the sound sources in a system. Better models contain six to eight lines with gain, bass, and treble controls on each channel. They also contain two to three phono and mic (microphone) inputs, a six-band graphic equalizer, bass and treble controls for the DJ mic, and a crossfader for mixing. They also include a master volume control, headphone volume, an LED scale, cueing, and other effects. Some mixers contain sophisticated sampling capability.

Amplifier

The primary purpose of an amp is to take low-voltage signals and make them stronger. Amplifier power is expressed in watts or, in stereo amplifiers, in watts per channel. The output load impedance is expressed in ohms, which relates to the maximum amount of load (from the speakers) that the amplifier is designed to drive. Large spaces, and some speakers, require high-powered amplifiers.

Speakers

Speakers reproduce specific audio frequencies. Speakers with crossovers that split the signal into two frequency ranges are called two-way systems. Crossovers that split the signal into three frequency ranges are three-way systems. Choose speakers that have a power rating 50 to 100 watts lower than your amplifier. It is recommended that you elevate your speakers to ear level using speaker stands.

Powered speakers

Lightweight, powered speaker cabinets feature multiple built-in amplifiers that produce great sound. These molded-plastic cabinets (i.e., JBL Eons, Mackie 450, or Yamaha) make perfect sense for mobile DJs. Each speaker has its own amplifier (for failure redundancy), is lightweight, quick and easy to set up, and more cabinets can be added (cascaded) as you book larger venues. To connect them, you just plug each cabinet into your mixer (via a balanced XLR mic cable) and then plug each cabinet into the wall outlet for power.

Bi-Amping

Consider bi-amping if you do a lot of jobs in large rooms such as gymnasiums, or if you want to add extra bass capability to your sound system. Bi-amping can be accomplished by purchasing one or two 15- or 18-inch subwoofers. You can either place both subwoofers next to each other in one corner of a room and couple them, or take one subwoofer with you and place it in the corner. If you bi-amp your system, the amplifier you purchase should have a power rating of 50 to 100 watts greater than the subwoofer. You will also need to purchase a crossover component for your sound system.

Microphones

The best wired microphones have dynamic or moving coil construction and are unidirectional. These types of microphones will pick up sound only from the person speaking directly into them. Wireless mics come with transmitters and are highly useful for blessings, toasts, speeches, and "working the crowd" on the dance floor. The best wireless mics are the diversity type although the non-diversity type is available. UHF mics usually work better than VHF mics because they transmit on less populated frequencies.

Headphone

A headphone allows the DJ to cue the music. The best type for mobile DJs is the single earpiece muff that covers the whole ear. Do not buy the open-air type or you may get feedback.

CD player(s)

A good dual-CD player is the mainstay of every good sound system. Many professional quality CD players come with pitch and speed control for mixing capability. Instant start and frame accuracy are the main features you will need. This type of player allows the DJ to cue up any downbeat and to start immediately. Dual transports are very popular with a detached remote control, as are single tabletop units. Full-feature models give the CD jock the ability to simulate scratching (as on a record album) and other turntable-like effects. Some even have protection from jarring so they will not skip even on the most crowded dance floors. Compact discs produce great sound quality.

Computer

Laptop or lunchbox style computers allow DJs to carry songs to spin music via MP3.

MP3 is a compression formula created to send music over the Internet. The standard for MP3 compression is 128 KBs (kilobytes per second). This allows for a 50 MB (megabyte) .wav file (a 3 minute stereo song) to be compressed to 9 percent (or 4.5 MB) of its original size to send or store it. For better quality playback you can compress at 160 KBs or 192 kps. The higher the number, the more hard disk space you will use. Compressing (ripping) your CDs to MP3 will allow for the storage of thousands of titles on your computer, making them accessible with the click of a mouse. The most popular software packages for DJs are DJ Power (www.djpower.com) and Visiosonic (www.visiosonic.com). The quality of the soundcard in your computer will be very important to the quality of the playback of these files.

MiniDisc

The MiniDisc (MD) is popular with some mobile DJs. Each disc can hold up to 74 minutes of digital audio. MDs include text and beats per minute (BPM) information for each disc and track. MDs enable you to pick a point midway through a track and divide it into two separate tracks so that you do not have to cue past unwanted intros. Some MD players have recording capability, and the recordings can be protected against accidental erasure.

Cassette deck

All cassette decks start, stop, play, pause, fast-forward, and rewind at the touch of a button. Better models include pitch control and a search feature that will automatically cue to the start of each song.

Turntables

These are becoming less important to the mobile disc jockey, although they are still used extensively by club DJs. The benefit of using a pro DJ turntable for mixing is that it features speed and pitch controls. An excellent turntable should start up within one-eighth to one-quarter turn and be a direct-drive model. Direct-drive turntables are superior for mixing and scratching. Felt antistatic mats are best for use with turntables. These types of mats provide the friction necessary to keep the record at a constant speed, allowing the DJ to slip the record in any direction.

Transporting Your Gear

There are a few different options for transporting your gear. If your system is small and compact, you may be able to transport it in your car. If your system is too large for a car, you can buy or rent a truck or van large enough to haul your equipment. Other options include buying a trailer to hitch up to your existing vehicle, or hiring a crew that has the ability to transport your equipment for you. Trailers carry lots of gear but are difficult to maneuver and park. Many DJs rent vans and take an insurance option supplied by the rental company.

Every DJ should own some type of hand truck to help roll the gear into a venue. *Rock n' Rollers* are probably the most popular device for mobiles and are available at nearly all DJ supply stores. *Magliners* (UPS's choice) are also excellent. Even if you transport the equipment yourself, you may want to consider hiring an assistant (roadie) to help you load, unload, set up, and break down your gear. It is a good idea to develop a checklist system for transporting your equipment. This will ensure that nothing is forgotten at home or after the job is over.

Set-Up Tips

You or your crew needs to arrive about one hour before the start time of the scheduled event to allow adequate time to set up and test the equipment. Upon arrival at a facility, the DJ should introduce himself or herself to the facility manager and client. Ensure that the location where you are setting up has properly grounded outlets and that a sturdy banquet table (with a tablecloth) has been prepared for you to place your music and equipment on.

Although most system components may be placed on a table, using a rack is better because it will make patching and transporting your system easier and it looks more professional. Some racks fit directly on a tabletop. If you wish to place your music there as well, you will need a table that is banquet size (approximately 2 ft × 8 ft). You can also use a soundstage for your racks and cases.

Always do a system sound check by testing each of the components at a very low volume level before the guests arrive. Once you are certain everything is operating properly, crank the system for a limited time to make sure you do not blow a circuit breaker once the party starts. Check for full fidelity.

If a piece of equipment is not working properly, it is essential to have backup equipment and a contingency plan. Consider hiring an on-call engineer who can be paged to come immediately to your location.

There are some major decisions to be made about sound and lighting gear when starting your mobile DJ service. Be an informed consumer. You can accomplish this by performing the necessary research prior to making a purchase. It is recommended that you talk to other local DJs, or log into DJ chat rooms on the Internet such as at www.djchat.com or www.prodj.com to examine all of your options.

Acknowledgment

A significant contribution to this chapter was made by Bernie Howard, a popular Chicagoland DJ since 1972. Bernie entertains at hundreds of different types of events annually and has been a radio personality and nightclub DJ. He is currently a Sales & Marketing Manager for Gemini Sound. Bernie writes for trade publications and Web sites, and has been crowned the "Highest Tech DJ on Earth." His sound system is second to none. You can contact him at Bernie@djberniehoward.com or visit his world at www.djberniehoward.com.

You are the music while
the music lasts.

T. S. ELIOT

5

Your Music Collection

*Are you the greatest entertainer in your market? Do you have
the best sound system and coolest lighting effects? None of that
matters unless you also have a quality, up-to-date music
collection that contains multiple genres from the 1940s to the
present. DJs who offer more have the right to charge a premium
for what they provide. Get the hint?*

CHUCK LEHNARD, OWNER
Spectrum Mobile DJ & Karaoke
Santa Rosa, California

Music Library Essentials

A complete music library contains cocktail, dinner, and dancing
music spanning the 1940s to the present. It includes every type of
music you will need for the events you perform. General categories
include oldies, Motown, big band, rock, Top 40, disco, country,
R&B, hip-hop, alternative, and new music.

It is important to read the trade publications to stay on
top of what music is currently popular and to purchase new
music every month. Your music collection needs to contain novelty, specialty, and ethnic music for use at appropriate functions.
www.djzone.net, www.mobilebeat.com, www.djtimes.com, and
www.partypros.com are examples of websites that contain excellent

music lists that include ballroom, country, ethnic, holiday, Top 200, and wedding songs and charts.

Here is a list of popular songs to consider as part of your music library:

Novelty, Specialty, and Participation Songs

ALLEY CAT	John Norris/Bent Fabric
BUNNY HOP	Ray Anthony
CHICKEN DANCE	Emeralds
CONGA	Miami Sound Machine
DADDY'S LITTLE GIRL	Al Martino
ELECTRIC BOOGIE	Marcia Griffiths
HAND JIVE	Johnny Otis Show
HAVAH NAGILA	Moshe Silberstein
HOKEY POKEY	Ray Anthony
HOT, HOT, HOT	Buster Poindexter
LIMBO ROCK	Chubby Checker
LOCOMOTION	Grand Funk Railroad
MACARENA	Los Del Rio (Bayside Boys Mix)
PARTY TRAIN	Gap Band
STROKIN'	Clarence Carter
TARANTELLA	John Norris
THE STROLL	Diamonds
TIME WARP	Rocky Horror Picture Show
WE ARE FAMILY	Sister Sledge
YMCA	Village People

Consider adding the following to your core music collection:

- Music for Jewish Bar/Bat Mitzvahs and weddings
- Music for various ethnic parties
- Christmas music
- Sound effects
- Grand entrance themes
- Show openers
- Marches
- Party closers

Music Services, Products, and Outlets

Whether you are just starting a collection, or would like a more affordable way to purchase new music, here are some helpful ways to stretch your music dollar.

The key to building a music library on a budget is to buy only what your audience demands, not every song you like. Bigger is not necessarily better! The music you need depends on the markets you serve. In other words, your music library will be defined by the types of parties and events you work and the demographics of the guests at those affairs.

It is less expensive to buy new music through a music service that is exclusively for DJs. When purchasing older music, be aware that there are many good compilations and collections that span all musical genres and decades. These collections can be found in most record stores, DJ-specific music suppliers, equipment catalogs, and on-line. Sometimes they are advertised on television.

New Music Services

(Current at publication time)

For a mobile to stay current with her/his tunes, there are a few excellent monthly and weekly CD services subscriptions for professional DJs. The best way to base a cost comparison is on a usable music per dollar ratio.

Promo Only is the largest and most popular music service for DJs and features 14 separate monthly genres. A Mainstream Radio subscription will guarantee a DJ nearly everything on the charts for $160 a year. For the mobile who does a lot of high school and Bar/Bat Mitzvah gigs, consider also adding a Rhythm Radio subscription, which includes more upbeat dance songs and popular rap. From there your selections will depend on your audiences' musical tastes. For jocks in metropolitan city areas, Urban Radio is a must. It features clean lyrics of all the R&B and rap songs. Country Radio is a great addition for DJs in rural areas or for those playing to country music–loving audiences. Modern Rock Radio provides current alternative rock songs, a genre that is needed for college and some middle and high school dances. For church events, *Promo Only* offers their Christian Series that features a hot, clean, and positive music mix. The rest of the formats are club and Latin oriented. There are three types of Latin: tropical (East Coast), regional (West Coast), and pop. For the club-oriented DJ who beat mixes, the club series CDs give you the extended dance mixes of songs. Mainstream club is pop remixes, followed by rhythm, alternative, import and underground.

Top Hits U.S.A., put out by *Radio Programming and Management*, offers adult contemporary, Top 40, and country by combining all their formats onto a weekly CD. The cost is $16.95 per week (you pay monthly at $79 a month) or $49.95 per month. Subscribers receive an extra CD every two months.

Prime Cuts is a service put out by *TM Century* and is another weekly service that combines all formats. Their price is slightly less than RPM, $39–$59 a month, depending on the length of the contract you sign. This service does not produce a combined disc every two months and does not offer an option for monthly service.

Nu Dance Traxx, Nu Life (Christian), Nu Music Traxx, Nu Country Traxx, and Knockout Hits are sold by *Entertainment Resources Group* for $15.95 each.

Remix Services

If you DJ at events or clubs where nonstop, high-energy, current dance music, and hipness will appeal, then consider looking into a remix service. These are not the original versions of songs but they can definitely add some freshness and style to these types of gigs.

Hot Tracks features exclusive dance remixes of popular songs on both CD and vinyl. They have four services that include Hot Tracks, Street Tracks, R&B, Roadkill!, and NRG for the '90s. Prices range from $14.95 to $26.95 per CD or vinyl.

X-MIX offers DJ remix compilations on CD and vinyl; Radio-Active, Dance, Urban, and Club Classics. Prices range from $15.95 to $26.95 for a two-CD set. They will soon be adding an Abducted series to their line featuring underground progressive house music.

Ultimix, one of the more popular remix services, offers Ultimix, Funkymix, Rampage!, Looking Back, and Rhythm & Scratch. Issues start at $22 and vary according to issue and CD or vinyl.

Music Video

With the price of DVD players constantly going down and the availability of LCD projectors for under $1000, for a minimal investment you can add music video services to your offering of entertainment options.

Promo Only, the largest CD DJ subscription company, offers music videos on DVD. Each DVD contains over 3 hours (over 40 videos), and there are many formats: Hot Video offers virtually every music video released every month; Club Video offers extended dance mixes made for mixing; Country Video; and Latin Video. Each of these are produced specifically for DJ use. Pop Mix and Dance Mix Video are premixed, continuous, and overlaid with special effects in the song transitions. Each subscription is $300 a year ($25 per DVD). The Hot Video Classics are available individually for $45 each.

ETV's DVD services include, Vital Dance, Vital Hits, and Vital Classics. Each is 2-hours long and contains new dance, new Top 40, and classics videos, in that order. Power Dance is 4 hours of dance songs with an emphasis on familiar older music. Each DVD is $79.

DJ Specialty Outlets

There are several DJ music specialty distributors that offer a variety of products. All of these can be found on the Internet or in a recent issue of your favorite DJ magazine. They all carry the major companies, but each one has its own specialties.

Disc Jockey Music Express catalog offers a variety of remix services, compilation CDs, and subscription services.

The Remix Warehouse catalog seems to specialize in remix services; the company also sells equipment.

The Source catalog has a variety of music services. This is the only West Coast–based company that can fulfill last minute orders for DJs on the East Coast.

Rotations mails out frequently updated listings of music for the pro mobile DJ. The company also has a Web site where it offers domestic and imported CD compilations in every music genre.

Bobby Morganstein Productions offers a series of CDs, which include Jewish, Latin, novelty, traditional, specialty, grand entrance, classic, cocktail, medley, big band, jingles, dinner, and Broadway

Figure 5.1 DJ Paul Binder and California Music Express. Courtesy *DJ Times* magazine.

music. The complete series of 13 CDs can be purchased for $255.00, or CDs can be purchased for $20.00 to $25.00 each. Music may also be purchased online through www.themobiledjhandbook.com.

To lower their costs, some DJs think nothing of buying a single copy of a CD and duplicating it for multiple systems, buying unlicensed compilations, or downloading music from the Internet onto MP3. Unauthorized duplication of copywritten material is illegal—and a shady business practice—that reflects poorly on the mobile DJ industry. The best way to save money and maintain your professional integrity is by carefully selecting the music you buy and from whom you buy it.

You can dramatically reduce your music costs by being a savvy shopper. This means you must research what is available and be proactive! Ask the managers at record stores you frequent for a DJ discount and search through their used CDs to supplement your library. Trade in the compact discs you do not use for new ones or for cash. Purchase compilations and collections whenever feasible. There are many ways to get more bang for your music buck.

Figure 5.2 Vendor booth at the International DJ Expo in Atlantic City, New Jersey. Courtesy *DJ Times* magazine.

Use the resources mentioned in this chapter, and watch your costs start shrinking.

Play Lists

Creating a play list of the music in your collection will prove to be very helpful on the job. It will assist you not only in making musical selections, but will also give your guests something to look at when they inquire about your music. The play list should be in alphabetical and categorical order. Place your play list in a clear plastic binder so it is easily readable and accessible. (*Note:* Some computer programs can do all the alphabetizing for you.) Your play list categories can include:

- Alphabetical by Title
- Alphabetical by Artist
- Hip-Hop/R&B/Rap
- New Music
- Disco

- Ballads
- Rock
- Oldies/Motown
- Big Band/Jazz/Swing
- Country
- Alternative
- Ethnic/Novelty

You may also want to add beats per minute (BPMs) to the information on your play list for your own use. This will help you to achieve great mixes at a glance by knowing what songs will work well together. BPM Publications publishes books that contain the beats per minute for almost every music category.

Latin Music

Over the past several years, Latin music has become increasingly popular. This musical genre includes salsa, merengue, Tejano, and charanga. The requests you may get for each type will depend on the nationality and geographic location of your audience.

Just like American music, Latin music is also constantly changing. It may not be necessary to have the very latest music from all artists, but it is a good idea to have at least a few songs that are popular from each year of the past decade. The quantity of Latin music you will need in your arsenal will depend on how often you get requests for it.

Latin classics usually work well both with older crowds and with younger audiences. Consider buying "Best of" compilations such as *Salsa Explosion*.

The Latin music industry's leading charts publication is *Radio y Música*. The online version (www.radioymusica.com) lists the top 10 songs per chart. The "Panel Bailable" (Danceable) and "Panel Tropical" lists the popular dance music. "Panel Tejano" lists the popular Tejano music as well. Another source for Latin music charts and information can be found at www.lamusica.com.

Swing Music

The most popular types of swing music are West Coast, East Coast, and Lindy. Each has its own tempo, structure, energy, and style.

Partners stay close to each other and dance in synchronization with West Coast Swing. Lifts and aerials are uncommon. The music is fairly sedate yet seductive. The style came into existence during the post-depression era when the popularity of big bands and ballrooms gave way to intimate jazz clubs.

People dance the Jitterbug to East Coast Swing, which is energy filled with lots of highflying action. The moves and variations are far more extravagant and spontaneous than with West Coast Swing. Dance partners continually rotate while executing basic patterns and variations including quick turns, Charleston kicks, and boogie walks. During the 1940s, couples made their moves to the sounds of the big bands that were popular at that time.

The "Lindy" was named after renowned aviator Charles Lindbergh and developed in the 1920s. The structure contains exaggerated body angles and patterns. The moves are extravagant and physically demanding. People do this dance to big band tunes.

Hey, Daddy-O/Mama-O (or should that be DJ-O)! There has been a reemergence of swing music, supported by groups and artists such as The Cherry Poppin' Daddies, The Royal Crown Revue, The Squirrel Nut Zippers, Bette Midler, Big Bad Voodoo Daddy, Tuxedo Junction, The Pointer Sisters, and The Manhattan Transfer.

Now go put on your zoot suit or sequined dress and remember, "It don't mean a thing if it ain't got that swing!"

To keep a lamp burning
we have to keep putting
oil in it.

MOTHER TERESA

6

Marketing and Advertising

Having a catchy business name, phone number, business cards, brochure, and website are all important marketing tools for a DJ. However, nothing is more important than direct marketing. This means that YOU are the show and the best means of advertising your company. Be a creative, passionate, professional performer and watch those word-of-mouth referrals grow without costing you a dime!

PETER GOLDSMITH, OWNER
Suburban Disc Jockey Productions
Bloomfield, Connecticut
Director of Marketing
www.themobiledjhandbook.com
on-line community

The Marketplace

The demand for mobile disc jockey entertainers is growing. People who would have hired a live band for previous functions are now hiring DJs for two main reasons. The first is cost. Mobile DJs are generally less expensive than bands. In addition, disc jockeys almost never take breaks, and clients get more music for their money.

The second reason is flexibility. A DJ provides all types of music. This includes big band, country, rock, oldies, hip-hop, alternative, and everything in between. Very few bands can competently cover such a wide range of music that suits every taste.

Communicate to potential clients that hiring a professional mobile DJ emcee far exceeds the entertainment that a band can provide. Illustrate how by showing your company DVD or video.

The market for mobile DJs is varied because not all of us pursue the same market. It is important to position your company toward the type of market that you will specialize in. It is necessary to gear all of your advertising and promotional literature toward this specific market or markets, depending upon your specialty.

A focused marketing campaign will yield far better results than a scattered one. There are several factors people consider when deciding upon a DJ. They will want to know about your experience with the type of affair they are holding. In addition, they may want to discuss the range of your music and whether or not you take requests (of course you do!). They may also make inquiries about your sound system. They may ask if you offer lighting and video effects or party-planning services.

Another factor to consider is your competition. It is necessary to do the research to find what types of jobs your competitors are booking and the rates they are charging.

Most successful mobile DJ services spend roughly 10 percent of their annual gross income on advertising and marketing. It is wise to commit at least this much during the first couple of years of operation, then decrease when referrals become significant. Although your referrals will grow as your reputation grows, it is not wise to rely solely upon word of mouth for new business.

Promotional Items

The use of promotional items is a good way to gain exposure and publicity for your mobile DJ service. Consider putting your company's name and logo on jackets, T-shirts, key tags, mugs, and refrigerator magnets. Form a company softball and/or volleyball team and play against radio and television stations. By doing this, you will very likely gain additional exposure through the media with whom you are competing.

Imagine giving the greatest performance of your life at a party and then having the guests say, "Great DJ, but I have no idea who she was!" Not a happy thought, right? It is important to visually identify your DJ service at every job using a banner.

Buy a tasteful, professionally made banner with your company name and logo. Affix it to the front of your sound system using Velcro or clips. The banner will identify your DJ service, serve as a decoration, and can hide wires and connections.

Business cards are your single most important marketing tool. Spending the extra money to have them look highly professional is well worth the expense. To make your business cards stand out you can have them printed with colored ink or on colored paper, or use gold or silver foil on shiny black stock instead of the standard white stock. You may want your business cards printed on extra-thick stock so they are heavier than most.

On a daily basis, keep this thought in mind: Every person you meet is a potential client!

If you are a part-time DJ, you can use contacts at your place of business to sell your disc jockey services. Give co-workers who are planning celebrations a note of congratulations with a business card tucked inside. Find out who hires the entertainment for the company holiday party and summer picnic and make sure they get cards as well.

Put a small stack of business cards on top of your speakers or on each guest table at every booking. Your clients' guests are the people who have seen you and can book you for their own affair, or refer you to others.

Another place to market your company is in the school systems. Call high school activity advisors. Proms, dances, and graduation exercises happen yearly. Contact nonprofit organizations that hold fundraising events. Introduce yourself and your company and offer them a special nonprofit discount. Send a couple of business cards and a company brochure.

Look for community bulletin boards at supermarkets, record shops, party-supply stores, or anywhere that will give your business card exposure. If you are active in an organization in your community, this is a great way to meet many new people. The time you devote to networking at organizations will come back to you multifold in the form of new business.

Always carry an ample supply of business cards with you. Distribute them to your friends, family members, and neighbors. Leave them every place you do business. Be sure to pass out a few cards to the banquet manager, photographer, and videographer when you are booked at a wedding, and ask for theirs too.

Keep them if you like the job they did! These folks can pass your business cards along to potential customers.

Digital Business Cards

The digital CD-ROM business card combines a conventional card with the data storage of a CD-ROM. Although shaped like a rectangle, the digital business card can be played in any standard CD-ROM drive or CD player that handles the 3-inch–size media. A prerecorded card usually contains business information on the front in full color and will hold 18 MB of uncompressed data and 40 MB with compression. Recordable digital business cards can be made in a standard CD or CD-ROM recorder and will hold up to 9 MB of data uncompressed and 12 MB with compression.

This high-tech media can be used as part of your DJ company's marketing materials. It is an excellent way to distribute brochures, photographs, audio and DVD, or video clips to your technologically savvy clientele. It can even be used to promote your Web site and include links to its on-line content.

Brochures and Cover Letters

Having a brochure to offer prospective clients is essential to your marketing and sales efforts. Desktop publishing has provided a breakthrough for small business owners. There are now preprinted brochure templates that can be purchased and used with the publishing program on a computer to create professional-looking promotional literature. You can also have your brochures professionally designed and printed on slick paper with multiple colored inks; however, be advised that there is a large cost difference between these two methods.

In the event that creating marketing materials is just not your forte, you may want to hire a professional to design your brochure. You can also purchase ready-made brochures in which you just add your company's specific information. Both of these types of brochures will cost more than creating one yourself, but the results will be professional and effective. It is not advisable or effective to invest money in promotional literature that looks amateurish in content and design.

The image presented to potential customers through your pro-motional literature needs to be polished to produce the greatest results. With brochures, perception is reality. It is my belief that hav-ing a slick, colorful brochure is well worth the expense—whatever its cost.

Statistics show that the most effective cover letters used for marketing contain the following qualities:

- They are personalized.
- They are two-sided and printed in Courier typeface.
- They contain headings and subheadings printed in blue.
- They are written in conversational language.
- The first side ends in the middle of a sentence.
- The margins are not justified.
- Key points are underlined.
- There is a P.S. at the bottom of the letter.

Effective brochures contain the following qualities:

- They consist of two to four colors.
- Pictures and/or graphics are included.
- There is a substantial amount of "white space" around the copy.
- They sell the benefits of your services, as well as allay the fears a client may have in hiring a DJ, and appeal to the emotions of the reader.

Consider creating a three-fold, 8½″ × 11″ brochure, or a two-fold, 16″ × 11″ brochure.

Direct Mail

Your direct-mail package should include a cover letter and a brochure, and in some cases, a coupon. It is important to know that the goal of direct mail is not to sell your service, but rather to prompt a response so that you can sell your service.

Some bridal publications offer direct-mail lists of engaged couples. These lists are often broken down by geographic region and can be highly effective for tapping into the lucrative wedding market.

DISC JOCKEY SERVICES

It's often said that music makes the party, and we at Paramount Entertainment could not agree more.

An experienced disc jockey can often be the difference in turning a good party into a great one. And great parties just happen to be what we specialize in.

Our disc jockey entertainers use audience participation to fuel their performances. By interacting with dance floor participants, and motivating standing on-lookers, our disc jockeys ensure the involvement of each and every guest while creating an enjoyable social atmosphere.

Each year our disc jockey personalities successfully host hundreds of events, in both the corporate and private markets. Combine this performance experience with our extensive music library, reliable mobile sound systems, and a professional image, and you've got what our clients like to call "ENTERTAINMENT EXCELLENCE".

Paramount's disc jockey services, however, don't just begin when your first guests arrive. In fact, our services start months earlier, with you and our event consultants discussing the details of your upcoming function.

Whether you wish to inform us of a music request, or simply confirm your evening's itinerary, our event consultants will assist you in all of your party planning decisions.

When the day of your event finally does arrive, you can trust that your Paramount Entertainment disc jockey will arrive on time, be appropriately dressed, and fully informed as to the evening's agenda. Nothing is left to chance.

Make the right choice ...
Trust your event to the party professionals at Paramount Entertainment.

40 VOGELL RD, UNIT 57 RICHMOND HILL, ONT. L4B 3N6 (905) 508-8530

PARAMOUNT

DISC JOCKEY SERVICES

Figure 6.1 Sample promotional brochure insert. Courtesy Paramount Entertainment.

Figure 6.2 Sample promotional brochure.

Let us turn your next party into a truly festive occasion.

Sound Performance is your 'one stop entertainment source' for the affair of a lifetime.

Select from a wide variety of affordable, top quality entertainment packages that include music spanning the decades. Our extensive record library features sounds for every taste from the 40's thru the 80's. Big Band, Disco/Dance, Oldies, Pop, Rock, Country, Ethnic, Traditional... and everything in between.

A sound for all seasons
State-of-the-art audio equipment is all we use, producing a quality of sound that is uncompromising and

unequaled. You'll not only hear the music — you'll feel the beat.

Trip the light fantastic.
A host of dazzling lighting effects are available to create a special mood for any theme — from the romantic to the fantastic. Mirror balls, strobes, traffic lights, revolving colors — even those popular bubble and fog machines.

Catch a rising star
Full color & sound videotaping of your event is also available — along with expert still photography by

highly experienced professionals. Both will be cherished for years to come as a treasured keepsake.

$50 off your first affair
We're confident that if we please you the first time, you'll call us again and again — and tell your friends about Sound Performance. But for this to happen, there has to be a *first* time. So as an added touch of "friendly persuasion" we're offering you a $50 discount on your first affair. In addition, the friends and business associates you recommend to us will also be extended this same first-time courtesy.

With Sound Performance you can relax...feel confident...and enjoy your own party!

Figure 6.2 (*Continued*)

Figure 6.3 Sample promotional brochure. Courtesy Mark's Rolling Dance Revue.

You may want to obtain lists that will be helpful toward your marketing efforts. Include the names of high schools and colleges, people getting married, and organizations and businesses that regularly put on socials.

"Two ingredients go into any good party – great food and great entertainment – Mark's Rolling Dance Revue exemplifies great entertainment."
FRED WITHEE,
GENERAL MANAGER
STORROWTON

"Five Stars ******"
ED & LUCY BUDZ,
BAR & BANQUET
MANAGERS
HAMPDEN COUNTRY
CLUB

"Being in the entertainment business, when Rock 102 needs its own entertainment, it looks to Mark's Rolling Dance Revue (M.R.D.R.). Whether it's a D.J. or Karaoke for a staff party or additional D.J. support for special events and promotions, M.R.D.R. always provides professional D.J.'s and exceeds our expectation. Rock 102 recommends Mark's Rolling Dance Revue to entertain at your next party."
WAQY-FM, "ROCK 102"

"The Only truly professional" D.J. company."
DONNA CLARKE
CATERING & SALES
MANAGER
GOODWIN HOTEL OF
HARTFORD

Exciting, entertaining, always professional. From weddings to corporate affairs, Mark's Rolling Dance Revue will make you party truly an event to remember.

Over a decade of experience in the mobile entertainment industry has taught us one thing: great parties don't just happen, they're the result of careful planning from the first to the last detail. Yet of all the decisions you must make, your choice of entertainment will have the greatest impact on your party's success.

Mark's Rolling Dance Revue is your guarantee of a good time. Since 1979, we have performed to over 1 million party attendees. Today, we are the largest professional Mobile D.J. service in the area, capable of accommodating over 35 functions in a single day!

The D.J. Service Professionals Recommend Most

Our professionalism both on and off the dance floor has earned Mark's Rolling Dance Revue a sterling reputation among the area's finest banquet facilities, photographers, and videographers. They know Mark's Rolling Dance Revue is the best insurance that a party will be a success. A complete list of industry references is available upon request.

Unforgettable Entertainers

People remember our parties because our D.J.'s really perform. While others simply stand, or worse, sit and mic the music, our D.J.s are on their feet entertaining, motivating, and keeping the party rolling. Ranging in age from 19 to 45 years young, our D.J.s offer a variety of unique personalities and diverse ethnic backgrounds to help you choose the right performers for your event.

Most have experience in professional radio, and/or are alumni of communications or broadcasting school. In addition, all D.J.'s are graduates of our affiliated training facility, Mass Mobile Disc Jockey School, the first and only vocational school of its kind licensed by the Mass. Department of Education. After graduation, new D.J.s must fulfill a six month apprenticeship before performing solo.

D.J.s are always appropriately dressed – formal attire for women and tuxedos for men.

Professional Party Sound

Mark's Rolling Dance Revue uses only high end commercial quality sound equipment (up to 10,000 watts). We can provide systems for groups as small as 25 people up to a crowd of 2,500!

Our equipment is set up on custom built stands to ensure a neat appearance with no unsightly wires. Traditionally, setup is completed 1 hour prior to the event. Backup equipment and personnel are always readily available.

Music Makes the Memories

Each D.J. carries a diverse music library of 5-10,000 selections on high quality CDs, ranging from the '30s to today's hottest sounds. Oldies, Rock, Dance, Country, Big Band and ethnic favorites are always on hand. Requests are welcome and line dances are encouraged. Your guests will relax and join in the fun as our DJs lead them through the Locomotion, Stroll, Electric Slide, Achy Breaky and newest dance craze.

For weddings, you'll find all the traditional music you require. Our D.J.s serve as full masters of ceremony. We handle bridal party introductions, cake cutting, the bouquet and garter loss, "Daddy's Little Girl," ethnic dances, pre-wedding meetings, and special requests and announcements.

Special Effects for every occasion

To add to the excitement, Mark's Rolling Dance Revue offers spectacular lighting equipment and special effects including fog and bubble machines, choreographed dancers and Laser Karaoke with giant video screens for singalongs.

For the Mitzvahs we offer 1 or 2 D.J.s, family introductions, Hora and Have Nagila, candlelight ceremony, and a variety of games, gifts, props, and prizes.

Wherever there's a party, you'll find Mark's Rolling Dance Revue. We offer quality entertainment for: *Weddings* *Jack & Jills* *Anniversaries* *Bar Mitzvahs* *Dances* *Class Reunions* *Retirement Parties* *Corporate Functions* *Christmas & New Year's Eve* *Graduations* *Communions* *Birthdays* *Picnics* *Roasts* *Banquets* *Fundraisers* *Karaoke Parties*

"Very professional and dependable services rendered."
LINDA SKOLE, OWNER
CHEZ JOSEF
AGAWAM, MA

"Top Notch Entertainment"
MARY NUGENTS,
MANAGER
MILL ON THE RIVER,
SOUTH WINDSOR, CT

"Distinctively different with unique professionalism."
STEVE ST. PETER,
BANQUET MANAGER
THE COLOSSEUM,
WEST SPRINGFIELD, MA

"Always a resounding success."
CONNIE O'BRIEN,
VICE PRESIDENT
WWLP-TV CHANNEL 22

"It takes great entertainment to entertain a radio station. And Mark's Rolling Dance Revue really entertains us."
MARK BERMAN
WHYN-AM/FM

Figure 6.3 (Continued)

Figure 6.4 Sample promotional postcard. Courtesy Denon & Doyle Disc Jockey Company.

In many cases you will only be able to get a general list, and additional research will be required before sending out your mailing. For example, you may have the name and address of a school or business but no contact name. Send your mailing to a specific person. This will greatly increase the chance that your information will be read and retained by the decision-making party. If possible, always follow up a mailing with a phone call within one week. You may want to say that you are calling to be sure the information you sent was received and to ask if he or she has any questions. Inquire about upcoming events. This person is your contact. Sell them on your DJ service.

With schools, send out separate mailings to the senior, junior, sophomore, and freshman class presidents. In addition, include the senior class advisor, student council president, student activities director, and junior and senior prom committee.

You may want to hold a contest at a bridal fair. Once you have a list of brides-to-be, send mailings to them. Also send mailings to engaged couples whose names are announced in the newspapers.

Other important direct-mail lists will include all hotels, VFW halls, banquet facilities, and such within the geographic area you serve. Folks seeking a mobile DJ entertainer will often call these places and ask for a referral. You want your information to go to the person who is in charge of bookings. Send direct mailings to past customers once or twice per year. They are an important source for referrals.

Free Publicity

Whether your DJ service is brand-new or already established in the marketplace, it is recommended that you seek out each and every opportunity to gain unpaid exposure in the media. Free publicity can be a valuable tool to raise visibility, boost sales, and help launch a new product or service. One important way to receive this publicity is by submitting newsworthy press releases to the media.

When the media does a story on your company, it's almost like getting a third-party endorsement. Articles, unlike advertisements, do not cost you a dime. Before you can send out a press release, develop your "angle"—your reason for writing the story.

Some ideas for finding a good angle include promoting an upcoming charitable job, staff expansion, a new division, or a location change. You may also consider writing an educational article such as "Choosing the best entertainment for your important event," or "Avoiding an entertainment nightmare at your wedding reception." Oftentimes it pays to be proactive. In other words, create a newsworthy event rather than waiting for one to happen. You might also contact publications and offer your services as a writer on topics within your area of expertise.

Define your target audience and obtain a media list at the library. Send the release to every newspaper, magazine, radio, and television station you think may be interested in the story. If you have a high-quality video of the subject of your release, submit this to TV stations with your press release.

Be sure to know each media deadline and the names of the persons to whom you should submit your release. When submitting to a newspaper, send your release to the editor of every section in which your story may be relevant.

It may be worth hiring an independent publicist if you do not trust your talent for writing a good press release. Not only can this type of individual create future ideas for your DJ business, but he or she will write the release and submit it to the media for you.

Here is the basic format for a press release:

- Type your release on 8½-inch × 11-inch letterhead with 1-inch margins on all sides, using only one side of the paper.
- In the upper left-hand corner type: FOR IMMEDIATE RELEASE; If it should be released on a certain date type: FOR RELEASE (followed by the MONTH and DAY).
- Date your release in the upper right-hand corner.
- Include a contact name and phone number underneath the release date.
- Keep the length to one page whenever possible. If you have more than one page, use a paper clip to hold the sheets together; do not staple them together. Place an identifying word or phrase in the upper left-hand corner of each page following the first. It is important not to carry over a paragraph to the following page.
- Begin each sentence with action-oriented verbs. Limit paragraphs to two or three sentences.
- At the end of the release type three hash marks, # # #, centered on the bottom of the page. This allows the reader to know your release has concluded.

Always stress the most important points first with the lesser points following. The first paragraph should answer the questions: who, what, when, where, and why. Proofread for spelling and grammatical errors. Ensure that your message is simple and clear enough to be understood by anyone.

When possible, include a couple of 3-inch × 5-inch or 5-inch × 7-inch color, action-oriented photos. On the back of the photos write the name of your DJ company, its address, phone number, website address, and the names of the individuals appearing. You may want to include a caption. The photographs should illustrate some important point made in the press release.

Sample Press Releases

Here are a few sample press releases from fictional mobile DJ services:

FOR IMMEDIATE RELEASE   Date
                                      Contact: Greta Garbanzo
                                             (000) 000-0000

CANNON FIRES ON TOWN GREEN

Sound Entertainment Mobile DJ Service will be shooting one ton of shredded paper from a cannon at the West Hartford Town Green on Saturday, April 15, at noon.

Sounds like quite a mess, right? Well, the town of West Hartford is using the cannon shot as a symbol to kick off its new recycling program. Mayor Carolyn Matthews will speak to the crowd about recycling and local television and radio personalities will make appearances throughout the day.

Pony rides, face-painting, and other activities for both young-sters and adults will be featured. The cheerleading squads from both Hall and Conard high schools will host a pep rally for the event.

The event begins at high noon and runs until 7:00 P.M. Admission is free, and food and drink will be available.

# # #

FOR IMMEDIATE RELEASE   Date
                                   Contact: Murphy McMullen
                                          (000) 000-0000

HOW TO CHOOSE A PROFESSIONAL MOBILE DISC JOCKEY

Choosing a truly professional mobile disc jockey to entertain your guests is perhaps the most important aspect of ensuring a successful event. Even when everything else is perfect, if the entertainment isn't fantastic, the whole event can be ruined.

How can you be sure to make the right decision? By being a smart entertainment shopper. You can hire a disc jockey you have already seen and liked, or you can ask someone whose opinion you trust for a referral. If neither of these choices is an option, there are always the Yellow Pages and the advertising in publications to which to turn.

When calling a mobile DJ service to obtain information and rates, do not judge value on price alone. If one company charges more than another, they may well be worth it. Anyone who has ever had an affair ruined due to an unprofessional or ill-prepared disc jockey can attest to this fact. You want to find an experienced DJ with an established reputation who has an extensive music library and state-of-the-art equipment.

Pay close attention to the company's professionalism over the phone. Here are the top 10 questions to ask:

1. Is this your full-time business?
2. Can I choose my own disc jockey?
3. What music will the DJ bring to my event?
4. What type of equipment do you provide?
5. How will my DJ be dressed?
6. Can the disc jockey teach popular dances?
7. Do you guarantee your services?
8. Can you provide references?
9. Do you offer party-planning services?
10. How much do your services cost?

When possible, visit the company's office to view a DVD or video of their DJ entertainers in action. At the very least, ask for two or three references and call them!

There are many professional mobile DJs to choose from in the marketplace. Make an informed decision based on becoming an educated consumer.

Murphy McMullen is president of In the Groove Mobile DJ Service and has been a professional mobile DJ for 15 years. For more information he can be reached at (000) 000-0000.

###

FOR IMMEDIATE RELEASE

Date

Contact: Neale Walton

(000) 000-0000

CALIFORNIA UNIVERSITY TO SPONSOR RETRO DANCE

California University is sponsoring a retro dance on Saturday, September 1, at 9:00 P.M. to benefit the school's cheerleading squad. Entertainment will be provided by Cream of the Crop Entertainment, a local disc jockey company featuring state-of-the-art sound and lighting equipment. '70s and '80s attire is encouraged. Tickets are $6.00 in advance and $8.00 at the door. Advance tickets may be purchased at the student union. Refreshments will be provided.

###

FOR IMMEDIATE RELEASE

Date

Contact: Roxanna White

(000) 000-0000

LOCAL DJ COMPANY EXPANDS STAFF

More Music Mobile Disc Jockey Service is proud to announce the addition of a talented new DJ to their staff: Bill Clintan. According to More Music president, Roxanna White, "Mr. Clintan brings several years of radio and mobile disc jockey experience to our company. We're very excited to have him on board."

Formed in 1993, More Music Mobile DJ Service has entertained at thousands of weddings, functions, and events throughout the area. Their experienced DJ personalities, extensive music library, and state-of-the-art equipment makes them one of the most sought after DJ services in Rhode Island.

#

Save all of the press you receive to impress your clients. You can laminate and frame articles for display in your office, or keep them in a photo album placed in your reception area.

Free publicity gained through the use of press releases can lead to increased profits for your business. Seek out and create opportunities to appear in the media. This will increase your credibility and the public's awareness of your mobile DJ service.

Advertising Sources

To be a leading force in the mobile DJ marketplace, you must be competitive with other DJ services. By carefully selecting the most beneficial media, you can expose your service to thousands of potential new clients.

Consider establishing an in-house advertising agency or using the services of a media buyer/planner. In either case, you can save 15 percent or more on all of the media you buy for your mobile DJ service. If you establish an advertising agency within your company, open up a separate business checking account in the agency's name, and print up separate stationary.

Newspapers and Publications

In almost every state there are weekly newspapers that have multiple-county editions. Advertising in these papers is usually cost effective. Place your ad under the "Entertainment" subtitle of the classified section.

A run of about 4 weeks gives you time to assess how well the ad is working. If, after this time period, there has been little or no response, it is time to change the ad or the media. If it is working, keep the ad running. Most media offer prepay and advertising agency discounts and additional discounts for long runs.

Your ad should bring in enough new business to pay for itself at least twice over. This is the best measurement of the ad's effectiveness.

If six different DJ services advertise in a local newspaper that has a circulation of 150,000, those six disc jockeys are locked into tight competition. If you put you ad into a smaller newspaper and you are the only one in it, you will get all of the leads from that paper. It is clear which is the more cost-effective media.

Publications to consider advertising in are bridal and entertainment guides, religious publications, community shoppers, and pennysavers. Newspapers and newsletters published by special interest groups can sometimes be effective sources for advertising.

Regardless of the type of business you are trying to secure through your advertising, lead the season by at least a couple of months. Advertising for Christmas parties starts in late October or early November. Run wedding reception advertisements in late March or early April.

Choose your advertising sources carefully and record the results. Each time a potential new client calls, inquire how they heard about your company.

Yellow Pages

This is an expensive but often necessary form of advertising. Most telephone books have specific "Disc Jockey" and "Entertainment Bureau" sections.

Many of your potential customers will go directly to the Yellow Pages when they are looking for a mobile DJ. Statistics show that consumers look first at the largest ads, ads with color, and companies whose names start at the beginning of the alphabet. You may want to keep these facts in mind when choosing a business name and when placing an ad in the Yellow Pages. If a display ad is not in your budget, perhaps you can consider a bold business listing that includes a super slogan, a fabulous logo, the use of color, or anything that makes your service stand apart from the competition.

Some disc jockey services prefer to have only a business listing. Their reasoning is that they believe people who use the Yellow Pages are only shopping for the best price, and this is not their target demographic.

Whatever method or methods of advertising and marketing you choose, consider them a test. Develop a system for monitoring the results of each area (cost versus return), and stay with what works best for your company.

Bridal Shows

If weddings are a mainstay of your business, then it may be important to secure a booth at every major bridal show within the geographic area you cover. Here are a few ideas that will ensure that your investment pays off:

1. If you have several disc jockeys who work for you, persuade a few of them to "work the booth" with you on a rotating basis. You can entice your DJs by letting them know that this is a great opportunity for them to get booked for a wedding. Instruct the DJs to talk to brides and grooms about their personal style of entertaining at wedding receptions.
2. Offer a 5 to 15 percent discount to anyone who books your services during the bridal show. Ask that they give you a deposit, and then give them a receipt. Mail their contract within a week.
3. Continuously show your company video.
4. Conduct a drawing for a free DJ. Have brides and grooms complete a coupon that includes their name, the name of their fiancé or fiancée, address, telephone number, and wedding date. This drawing will give you an excellent mailing list to use after the bridal show.
5. If possible, set up one of your sound and lighting systems and have your DJs play music. This is especially effective when conducted from a stage.

Trade Deals

The value of a good trade deal cannot be underestimated. If used wisely, these deals can save you thousands of dollars of out-of-pocket expenses.

Think about the areas of greatest expense to your business and, when possible, offer to barter your service for a product or service from suppliers. Consider doing a trade deal for radio advertising. Perhaps a supplier does not want what you are offering, but they do want something else. You may be able to barter with yet another business that *does* have what the supplier wants. Look in the Yellow Pages under "Barter Services" for barter houses that specialize in setting up such exchanges.

If you decide to use the services of a barter company, obtain references and thoroughly research their credibility. The investigation is necessary, otherwise you may find yourself in the position of doing jobs to accumulate points and then have no way to cash them in for products or services you want or need.

In any event, do not use barter to an extreme. Cash flow is the lifeblood of every business, and if you do too much trading, you could be hurting your business.

Telemarketing

Telemarketing to prequalified prospects can be an invaluable tool for initiating new business. Put simply, it is marketing your services over the telephone, and it can be an integral part of your overall, on-going marketing campaign.

You will need to do the telemarketing yourself or hire someone to do it for you. Sales are an important aspect of being in business for yourself; getting used to and comfortable with the process is necessary.

The goal of your telemarketing effort is to get people to hire you, or to request more information about your services. To be effective, you must begin with a targeted list of prospects and prepare a script of what you will say. Then, when you are ready, pick up the phone and make those calls!

Your target prospects may include nonprofit organizations, corporate activities directors, brides to be (engagements are listed in the state and local newspapers), high school and college activity directors, newly opening retail stores (check with the chamber of commerce), country clubs, party equipment rental stores, restaurants, night clubs, function halls, river cruise boats, VFWs, hotels, and so on.

You can find the information you need to develop these lists in the Yellow Pages, newspapers, and business journals. Some bridal publications offer lists of engaged couples.

With schools, the contact names will change every year, so keep abreast of who is currently in charge. In high schools, the student advisor or activities director can give you the names of the freshman, sophomore, junior, and senior class officers, as well as the president of the student council. With colleges, many student organizations hire entertainment. Contact the student activities office and find out the names of the campus organizations, and do not forget the fraternities, sororities, and minority groups. The alumni organizations hold class reunions every year.

Developing key contacts will significantly assist your telemarketing efforts. Take the time to meet with key people in person. Consider offering them a financial or other personal incentive to provide you with referrals. Then, when you call them, they are more likely to be cooperative with you. After you have worked out a mutually beneficial arrangement, do not wait for your contacts to call you. The responsibility lies with you to touch base with them every month. The ideal situation will be one in which you develop a long-term relationship. In this instance, the initial time and effort you have expended will pay off in numerous on-going referrals.

Perhaps the most important prospect list of all is that of your past customers. Call or write to them twice yearly to inquire about their future occasions and any referrals they may have for you. You may want to offer past customers a financial incentive for assisting you.

The script

A truly effective script should not sound like it is being read. Its tone should be friendly, conversational, compelling, and brief. Target the script toward the type of event the prospect may be planning. Here is an example:

> *"Hi, Mrs. Smith, this is John Hyper from Platinum DJ Service. We perform DJ, lighting, and video shows all over New England. The reason I'm calling is to see if you would like some information on our services with no obligation. (Get the correct spelling of the prospect's name and his or her address.) Thank you. By the way, are you planning an event this year? Great, what kind of event is it? Well, I'll send you out the information right away, but while we're on the phone I'd like to let you know that we specialize in (name whatever event the prospect just mentioned)..."*

Getting appointments

With most wedding and Bat/Bar Mitzvah bookings, you will probably need to set up an appointment to close the sale. In this case,
gear your telemarketing efforts toward getting the potential client
to meet with you, not toward closing the sale over the phone.
Once you meet the interested party, you can close the sale in
the same manner you usually use over the phone. You will also
have the added assistance of your company video and marketing
materials.

With wedding prospects, make an appointment in which both
the bride and groom can attend. If their parents are instrumental to the arrangements (such as paying for the reception), include
them as well. When making an appointment with Bar/Bat Mitzvah
prospects, both parents and the youngster will need to attend.

If your telemarketing goal is to get someone to meet with you,
your script needs to be altered from the phone-closing script. The
content remains the same, but the close is different:

> *"The reason I'm calling is to see if we could get together for a few minutes
> to talk about your wedding reception plans. I'd like to show you our short
> video and share some ideas with you that will ensure your reception's
> success. When would be a better time for us to meet; Tuesday evening at
> 6:00 P.M. or Thursday evening at 7:30 P.M.?"*

Producing a Website

Planning is the most important aspect of producing a website. The
purpose of your site should be for the visitors to want to hire you or
at least to request a personal call or appointment. Make it easy for
them!

All websites are found via their Universal Resource Locator
(URL). This is the "street" address of your Internet website. Your
main (index) page should be simple and enticing and include links
to your other pages.

When producing a website, the primary areas to consider are
navigation (how the client will get from one page to the next);
content (the information in the site itself); maintenance (how you
will keep the information updated); mechanics (where your site will
appear on the Internet and how much it will cost); and marketing
(how will people find out about your site).

Continuity is important, so if you have developed a certain look or style for your DJ service, develop a website that is compatible and reflects your image.

Features

There are a lot of technically sophisticated features that can be incorporated into a website. These include options such as the ability to search your whole music library on-line by title, artist, and style and to process requests. You can use animated graphics, buttons and menus, banners, bullets, backgrounds, animation, tables, frames, sound and video. Using all of the latest multimedia features is like printing a 3D, interactive brochure. On-line payment processing could also be added that automatically processes payments from your clients.

Navigation

Your navigation can be set up in many different ways, from the simple link to a graphic link. Good navigation is the key to giving a client the information they want fast. Your home page (opening page) should have navigation set up with links to pages that are created for the different divisions or services your company offers. On each of those pages, create subdivisions that quantify your services into separate sections.

Consider including pages for your music library database, your company profile, services and products, DJ biographies, rates, equipment listing, referrals, streaming video of events where you have performed, employment opportunities, and a client information form. A links page with local vendors is a great way to network. Place a link to their Web site in exchange for your website's URL being included in theirs. Of course you will also want to include your company's address, phone and fax numbers, and E-mail address along with your copyright at the bottom of each of these pages.

Cost

There are several costs associated with running a website. They include site development, hosting, and maintenance. The cost of producing a website ranges from a few hundred to several thousand dollars depending upon the complexity of the site and the features it has available.

Search Engines

The secret to search-engine placement is in the hands of the person who creates your site. Hidden within the code for each page of your site are a couple of lines called META tags. These include a description of what your company is and what it does, along with keywords associated with your company such as your location and the events at which you perform. The person creating your site needs limited computer skills to write META tags into your code.

After you have created your website and your META tags are in place, register it with the larger search engines so that people can locate you. When submitting information, be sure the description of your business includes the territory you cover along with the types of events at which your company performs. For a listing of some of the biggest search engines, go to http://www.searchenginewatch.com.

Customer Service

People expect E-mails to be answered within hours; within a day at the most. E-mail should be treated more like a phone call than like regular mail. If your company cannot handle inquiries quickly, you might want to hold off on having a website until you can. Otherwise, you will be hurting your reputation.

Maintenance

Keep your site current by revising existing Web documents or creating new ones. You may do this yourself, it could be included in your hosting package, or it may be outsourced, depending on your budget, time, and skills.

Another important aspect of site maintenance is checking your E-mail. At a minimum, check it once daily and respond promptly. If you travel you need to arrange remote access or have an associate check it for you.

Mechanics

Your website consists of a collection of files. Each Web page and graphic image is a file. To be accessible to the Internet and the World Wide Web these files need to be stored on a Web server. A Web server is simply a computer that is connected to the Internet. Your Web pages will be stored on the Web server's hard disk so that they are always available.

Most companies rent space on the Web server of an Internet hostprovider. These are companies that own Web servers with direct high-speed Internet connections. An access provider's system connects you to the Internet through a modem connection. Their system gives you optimal speed for two-way Internet connections. Host providers are optimized for serving files to those connected.

Some Internet access providers still provide Web-hosting services, but the trend has been to go with companies that specialize in hosting. There is also a difference in the services offered. Your Internet host provider typically handles domain name registration. To learn if the name you want is available, go to www.internic.net.

Marketing

To ensure the greatest success of your site, you will need to actively market it. Be sure to place your website address on your business cards, letterhead, brochures, pens, company vehicles, T-shirts, Yellow Page, and other advertisements, and on your business answering machine. Give clients your website address during telephone contact calls or during in-person consultations. Place it on Web directories and search engines. Have your home page linked to other complementary pages on the Web such as photography and video services with which you are affiliated. Research other mobiles' well-produced websites—the ones that quickly show up using a search engine.

Almost all the major disc jockey associations have Web sites that include member directories. Many will provide you with a free link as a benefit of membership. Another option is to pay for link sites that excel in marketing your company to prospective clients for very inexpensive fees. Other websites can be found that offer free listings that include a link. One such site is www.partypros.com.

The Web is in a constant state of change and improvement. Almost daily there are new technologies going on-line that give you the opportunity to improve your site. Be on the lookout for other sites you like, and keep track of them. With a good website that is properly marketed, you should expect to get qualified sales leads that will provide bookings and revenue for your mobile business.

Clubs and Organizations

Marketing to nonprofit, religious, social, and professional organizations and clubs is an excellent way to grow your mobile business. Join as many of these as you can afford and have the time to attend their gatherings. Talk to people about their work and yours. www.guidestar.org and www.idealist.org are two excellent on-line sources for locating national chapters—which usually have links to local chapters. Your local Yellow Pages will contain an extensive listing as well.

One way to gain professional recognition as a mobile disc jockey is to join some of these organizations and become an active volunteer. Your efforts will be rewarded when they hire you for their events. In addition, by networking with other club or organizational members and handing out your business cards, you could wind up being their DJ of choice for important life events. Both for-profit and nonprofit organizations and clubs can be great sources of referrals to mobile entertainers.

Schools

For a high-energy, interactive DJ with an up-to-date contemporary hit radio (CHR) music collection, the middle and high school markets can be very lucrative.

Although student groups change yearly, the school's activities director position is generally a bit more stable. He or she will be more likely to recommend you if you provide assurance that you do not play songs with questionable or offensive lyrics, even if requested.

Effective marketing to schools requires the use of contemporary, colorful materials that have pizzazz. They should contain color photographs of you and your sound system and lighting. Teenagers tend to be very impressed by large speakers with subwoofers and nightclub style lighting effects.

An effective tool for marketing yourself at an event is to be friendly and fun, to take requests, and give away CDs. Your local CHR radio station may be willing to give you boxes of CDs for free.

It is best to begin your marketing to schools at the beginning of September and again in January. You may wish to emphasize your experience with homecoming, after-the-game, winter formal, spring fling, prom, and graduation dance events.

You can obtain a listing of schools in your area, including addresses and phone numbers, from your local libraries, state Department of Education, or Yellow Pages directory.

Mobile Mitzvah Marketing

Successful marketing of your entertainment services is especially important in the Mitzvah realm because the income potential is so high. Even for a basic Bar/Bat Mitzvah, the revenue is significantly greater than with other types of parties and events. Prices can range from $500 to $5000 or more depending on lighting, staging, the number of DJs and dancers provided, and other add-ons such as Karaoke, magicians, face painters, carnival games, and so forth.

Most people do not use a phone book to shop for Mitzvah DJs. A marketing-savvy Mitzvah mobile must learn the locations of synagogues within his/her area and whenever possible, establish rapport with their rabbis. Getting into the community generally requires advertising in Jewish newspapers and periodicals as well as temple and Hebrew school publications. For exposure, you may even consider doing a free 1-hour party for a holiday such as Hanukah to gain exposure.

Once you are firmly established, marketing becomes less costly as the need to advertise diminishes and your word-of-mouth referrals grow.

You see things; and you say, "Why?" but I dream things that never were; and I say, "Why not?"

GEORGE BERNARD SHAW

7

Party Possibilities

Quality will take you much further than quantity in the long run. If you are ethical, honest, and quality conscious in your business, you will gain referrals. People who want you are generally willing to pay more for you. Do not try to be a jack-of-all-trades and master of none.

"PARADISE" MIKE ALEXANDER, OWNER
Paradise Disc Jockey & Entertainment Company
Orcutt, California

Weekend and Weekday Events

There are an abundance of private, corporate, and public events that are ideal bookings for a mobile DJ. Depending upon your personal skills and the skills of those who work for you, some or all of these will be right for your company.

American mobile DJs are exempt from paying licensing fees, provided that they play only for private parties or at facilities that are licensed by organizations such as The American Society of Composers, Authors, and Publishers (ASCAP) and Broadcast Music, Inc. (BMI). These organizations protect the rights of writers, performers, and publishers of recorded music. They ensure that their members receive proper compensation for their work. Radio and television stations and public entities that provide musical entertainment must pay licensing fees. It may be wise to include a clause in your client contract that legally covers you regarding this issue.

103

Here are some of the booking possibilities that exist in the marketplace:

- Weddings
- Corporate events
- Cruises
- Bowling alleys
- Proms
- Graduations
- College dances
- Holiday parties
- Nightclubs and bars
- Retirement parties
- House parties
- Karaoke
- Birthday parties
- Theme nights
- Grand openings
- Block parties
- Picnics
- Fashion shows
- Bar/Bat Mitzvahs
- Fund raisers
- Roller skating rinks
- Socials
- Conventions
- Jr. and Sr. High School dances
- Clubs and organizations
- Reunions
- Carnivals
- Restaurants
- Banquets
- Teen dances
- Singles dances
- Pool parties
- Anniversary parties
- Cocktail parties
- Fairs
- Rallies

Read the newspapers to find out about upcoming events, then contact the person in charge and explain how a disc jockey entertainer can enhance the occasion. Convince this person to hire your company.

Wedding Receptions

A wedding is the most important single event in a bride and groom's life. Treat each couple's reception as if it were your own and provide them with outstanding planning and entertainment services.

After you are hired, send the couple a package that includes a cover letter, song request list, and music list and bridal party introduction sheets. Request that the sheets be sent back to you two weeks prior to the wedding so you can review them personally, and with the couple, prior to the reception.

Certain key questions you want answered ahead of time are, Do you want the cake cutting to be "nice" or "naughty"? Are there any artists or songs you absolutely do not want played—even if they are requested by a guest? What type of interaction are you looking for at your event—high energy or low key? Do you have a liaison you would like me to deal with during the reception?

In your cover letter, ask the couple to write down their favorite songs and those that their guests are most likely to enjoy. Communicate to the clients that they may select as many songs as they wish, but they should keep in mind that there will only be time for 30 to 50 of the most danceable selections after the meal.

The bridal party introduction sheets that they send back will provide you with the names of the members of the bridal party and the order in which they should be introduced. It will also tell you what song the bride and groom want for their first dance, as well as for the father-and-daughter and mother-and-son dances. Additional information will include options such as the bouquet toss, garter toss, and cake-cutting ceremonies, and the Dollar Dance. These sheets will tell you when the bride and groom would like these events to happen during the reception. In addition, the sheet will inform you as to whether you will be provided with a meal at the reception.

At the party, program your music based on client selections, on-the-spot guest requests, and your own sense of what will work best.

At the reception, line up the bridal party for their introductions. Ask the bride if there are any changes to the information sheet she completed. Explain to the bridal party the cue on which each couple will enter the room. Go over each name for correct pronunciation and write out difficult ones phonetically. It is vitally important that you pronounce each name correctly!

It is your job to guarantee that the reception runs smoothly. Upon arrival at a wedding reception, speak with the banquet manager as well as the photographer and/or videographer to discuss with them the order and the timing of the events during the reception. It is important that your efforts are completely coordinated with theirs.

The following is a popular wedding reception itinerary. You and the client may customize the format as desired.

Guest arrival

Play cocktail music as the guests enter the facility and while pictures are being taken.

Bridal party introductions

Never announce any event without first warning the bride and groom, to ensure that they are ready. As you are announcing the bridal party, play instrumental background music until you are ready to announce the bride and groom. Then play the traditional *Wedding March* reprise or high-energy music.

Bride and groom first dance

Play the selected song.

OR blessing and toast

Do not play any music for the blessing or toast. Provide a wireless microphone to the person speaking. Prior to announcing the bridal party, instruct the person doing the toast to hand the microphone to the person doing the blessing when he or she is through. Make sure that the mic is turned on ahead of time with a fresh battery so that all you need to do is turn up the volume.

Bridal party dance

Play the selected song.

Bride and father dance

Play the selected song.

Groom and mother dance

Play the selected song.

Dinner

Play a mixture of instrumentals and vocals tailored to the guests' tastes. Right after dessert pick up the tempo to begin the dancing.

Cake cutting

Play low-key background music, or the traditional sing-along *The Bride Cuts the Cake*.

Bouquet toss

Announce that it's time for the bouquet and garter ceremonies. Ask for all the single women to come up to the dance floor. Play a female-oriented song. Position the bride directly in front of the ladies with her back to them. Use exciting or humorous patter to make comments on the event. Lead the guests in a countdown starting with five and ending with one. At the completion of the countdown, the bride will throw the bouquet.

Garter toss

Have a chair placed on the dance floor and ask the bride to take a seat. Encourage the guests to heckle the groom with words like "higher" as he removes the garter. *The Stripper* or a suggestive tune usually goes over well for the garter removal and toss.

After the groom has removed the garter, have the chair removed. Ask all of the single men to join the groom on the dance floor. The groom stands where the bride stood and tosses the garter in the same way the bride tossed her bouquet. Next, the guy who caught the garter places it on the leg of the gal who caught the bouquet. Play upbeat, sports-oriented, or dramatic music, and do a play-by-play description of what is happening similar to a sportscaster's commentary. If the crowd appears a little risqué, have the woman put the garter on the man after he has put it on her.

Anniversary dance (optional)

See "The Interactive DJ" in Chapter 3 for a description of the dance.

Conga line (optional)

Ask everyone on the dance floor to put their hands in the air and form a line behind the bride and groom. Play a conga-line favorite.

Dancing

Play a mixture of music that will appeal to all the age groups attending the reception. This usually includes, but is not limited to, oldies, ballads, big band, disco, Motown, and current dance music. The guests usually also enjoy being taught and leading participation or ethnic dances. Be sure to play requests from the client's song list, requests from guests, and your own favorites to keep the party going.

Bridal Party Introductions

Here is a standard bridal party introduction format:

Good afternoon/evening ladies and gentlemen, may I have your attention please? It's great to see everyone here at Tommy and Pamela's wedding reception. My name is Ellen, from Perfect DJ Corporation, and I'll be your emcee and DJ entertainer for the rest of the reception.

Right now it gives me great pleasure to introduce some very special people. Please give a warm round of applause for the parents of our bride, Mr. and Mrs. Richard Burton. Please continue applauding for the parents of the groom, Mr. and Mrs. Leonard Skynard. And now, ladies and gentlemen, the members of our wedding party. Starting things off, let's hear it for our bridesmaid Ginger Grant, escorted by usher Gilligan Smith [REPEAT FOR ALL BRIDESMAIDS AND USHERS]. Let's give a nice round of applause to our flower girl, Grace Slick, escorted by the ring bearer, David Crosby. Next, here come two VIPs in our wedding party. Put your hands together for our maid/matron of honor Diana Ross, escorted by best man Michael Jackson. And now, ladies and gentlemen, would you please rise. It is my honor and privilege to introduce to you for the first time in public as husband and wife, Mr. and Mrs. Tom Anderson. For their first dance together as husband and wife, Tommy and Pamela have chosen (fill in the blank). Ladies and gentlemen, let's give another round of applause for our newlyweds.

See sample wedding party introduction sheet and wedding reception questionnaire in Chapter 11.

Bar/Bat Mitzvahs

When a Jewish boy (Bar Mitzvah) or girl (Bat or Bas Mitzvah) turns 13, a religious service is held at his or her synagogue, usually on a Saturday morning. A celebration or Simcha follows the service. It is at the celebration where the mobile entertainer comes in.

There are three types of Judaism: Orthodox, Conservative, and Reform. Each has different customs and needs that you must have knowledge of in order to perform adequately at Bar/Bat Mitzvah receptions. It may be helpful to contact all three types of synagogues, and discuss with a rabbi the customs and practices that are important for you to know and understand.

It is not necessary to either know or to practice Judaism in order to sell and conduct these celebrations, but it is important to be aware of and sensitive to Jewish customs. For example, when you are doing a follow-up with a Conservative or Orthodox Jewish client, never call between Friday at sundown to Saturday at sundown because this is the Sabbath, a holy time. If the party is booked for a Saturday afternoon, this may give you some insight into how religiously liberal the client is and, generally, would be acceptable to call the client during the above time period.

Most mobile DJ services charge a premium price for a Bar/Bat Mitzvah celebration. Temple coordinators can be a great referral source into this market. Booking this type of event usually requires an in-person consultation. Both parents (or at least the mother) and the youngster being Bar or Bat Mitzvahed should be in attendance at the consultation. The celebrant will become your best salesperson of add-ons.

After booking a job, send out a "Bar/Bat Mitzvah Reception Questionnaire" to the client. This sheet allows the client to designate who will light candles and in what order.

Cocktail Hour

Prior to the party there is typically a cocktail hour, which is usually held outside of the main room in the foyer. If the adults prefer to be alone, keep the children busy with games and dancing in the main room while a second DJ plays music for the adults.

Grand Entrance

As the guests enter the main room, play upbeat music with your lights going, while your dancers/party prompters are on the dance floor urging the guests to join in. After a few minutes, introduce the Mitzvah with some high-energy music that is related to the theme of the celebration/decorations. The Mitzvah may simply walk into the room waving to the guests, enter in grand style of some sort, be escorted by your dancers, or be carried in by family members. Crank up the energy by having everyone clapping their hands and cheering for her/him.

Candle-Lighting Ceremony

The candle-lighting ceremony is usually conducted after the cocktail hour. Ask all of the guests to be seated so that the candles (13 or more) may be lit by the family and friends of the hosting family. The candles are lit to honor the person being Bar/Bat Mitzvahed. This celebration usually begins with the grandparents and ends with the Bar/Bat Mitzvah celebrant. Close friends and relatives are called to light candles as well.

Jewish music is played during the candle lighting. This often includes songs such as *Tradition; Sunrise, Sunset;* and *To Life* from the musical *Fiddler on the Roof. Simon Tov* and *Mazel Tov* are also usually played. The music may include more contemporary songs as well.

The DJ invites family and friends to come up and light a candle. There are times when the celebrant will read a poem and introduce the guests. After the last candle has been lit by the guest of honor, you ask the guests to reflect on his or her future. After a few moments he/she blows out the candles.

Hora

Next, invite all of the guests to join in a hora. A circle is made on the dance floor around the immediate family. The celebrant and the younger brothers and sisters are lifted in a chair. The traditional music played for this dance is *Havah Nagila*. The hora is the traditional Jewish dance done after the introduction. Announce, "Ladies and Gentlemen we want to start (Mitzvah's name) special night off right, so at this time everyone is invited onto the dance floor for the hora." Start the song, turn on your lighting effects, bring out your dancers, and instruct the guests about what to do. Ask them to form a circle or circles, holding hands with the person next to them. The outside circle moves to the left, and the inside circle moves to the right. You may want to pass out party props for use in the hora.

The Motzi and the Kiddish

After the hora, ask everyone to be seated and call a predesignated person to do the blessing over the bread and wine. This ritual is called the *Motzi* and the *Kiddish*. The Motzi is the blessing over the challah and the Kiddish is the blessing over the wine. Sometimes the client may want to have a *Havdalah* service. This is a traditional prayer and singing that signifies the end of *Shabbot*, the holy day.

Figure 7.1 Party props from Sherman Specialty Company. Courtesy *DJ Times*.

It is done a short while after sundown. Typically, the Mitzvah will do the prayer(s) alone or with family members or a friend or cousin.

Lunch/Dinner

During the meal it is standard to play more of the adults' requests. Announce to the youngsters that their turn is coming up soon. Generally they will be finished eating first, so have some games planned for them during mealtime.

After the guests are done eating, be ready to entertain the adults, teenagers, and children. This is accomplished through a mix of music and games that will appeal to all age groups. Many Bar/Bat Mitzvahs are booked with two DJ Entertainers or one emcee and one disc jockey. In either scenario, one DJ acts as the emcee/party host while the other programs the music and technically controls the music and lighting systems. Dancers, party assistants, a magician, a face painter, a balloonist, big-screen video, and Karaoke sing-along are also add-ons that are very common to this type of event.

High-energy dancers add to the entertainment value of the reception and can also assist the emcee by handing out props and prizes during games.

Dessert

During dessert some of the youngsters may want to present a memory glass to the guest of honor. In this ritual they gather up mementos from the party and put them in a glass with melted wax on the top.

Closing

Near the end of the affair, gather the guests in a circle on the dance floor. Play a sentimental favorite that honors friends and family.

After Party

Sometimes there is a separate party for the youngsters after the main reception. If your client wishes to book this add-on, the music played will be solely for the youngsters.

There are some excellent videos on the market that will show you how to conduct Bar/Bat Mitzvahs. There are also compact disc sets of music specifically designed for this event that are available. Consult the marketplace section of the *DJ Times* and *Mobile Beat* magazines, as well as DJ catalogs and Internet sites, to find out what is offered.

The Corporate Market

Within the corporate market there are several types of clients. These include:

- Companies. From "Mom and Pop" shops to Fortune 500s.
- Entertainment agencies. They book bands, mobile disc jockeys, Karaoke, etc.
- Nonprofit organizations. They hold special event fundraisers.
- Event planners. They usually provide full-service convention and conference planning services.
- Convention sites, hotels, resorts. Most of these act as referral sources for entertainment; some act as booking agents.

Figure 7.2 Man, these are the funniest looking hula-hoops I've ever seen! Courtesy *Mobile Beat*.

Holiday Party Tips

Christmas/Hanukah

- Spice up your tuxedo by wearing a red or green cummerbund, vest, and bow tie
- Read holiday trivia during cocktail hour and give away prizes to guests having the correct answers
- Gear your music to the primary religious/racial group to whom you are playing

Valentine's Day

- Play lots of slow tunes
- Hand out some hearts

St. Patrick's Day

- Play lots of Irish music
- Wear something green
- Hand out some shamrocks
- Tell occasional Irish limericks

Figure 7.3 DJ Randi Rae.

Fourth of July

- Wear patriotic colors
- Play pro-America music but do not overdo it
- Play U.S.-trivia games

Pool Parties

- Host raft and relay races, treasure dives, water balloon and hula-hoop contests
- Play Caribbean, Hawaiian, reggae, and beach-oriented songs

Halloween

- Wear a costume
- Host a costume contest
- Play the Halloween music people have come to expect

The achievements of an organization are the results of the combined effort of each individual.

VINCE LOMBARDI

8

Multi-System and Full-Service Operations

Going Multi-System is like instantly giving birth to quintuplets. With patience, research, and a lot of hard work it can be just as rewarding. If not, you will soon be wishing for the days of bachelorhood.

BRIAN DOYLE, CO-OWNER
Denon & Doyle Disc Jockey Company
Concord, California

Hiring Mobile DJs and Dancers

Disc Jockeys

As your bookings increase to a point where you alone can no longer fulfill them, you will need to consider hiring other DJs to work for your company. You can either match the skills of the DJs you hire to the type of events your company performs or expand the markets you serve by hiring DJs with skills and interests different from your own or who are from different gender, ethnic, or racial backgrounds.

You can hire DJs as employees or as independent contractors. With employees you will be able to assert a higher level of control but your costs will be greater. With independent contractors you will have less administrative and financial responsibility; however, your management options are limited. You can produce a statement of your company's quality and other types of standards necessary for independent contractors who work with you. It is up to each person to comply or not. The reader is strongly urged to become educated on the precise distinctions between independent contractors and employees prior to making a decision on this subject. There are numerous trade publication and website articles on this topic that are available. It is also wise to consult with an accountant and attorney to gain awareness of particular state and federal regulations.

Whether hiring an employee or a contracted worker, you will want to ensure that a DJ's performance abilities, equipment, and music meet your standards.

Recruiting

To recruit other mobile DJ entertainers it is a good idea to place ads in your local area entertainment publications and newspapers. You can also place help-wanted flyers in DJ/sound equipment stores, music shops, and on the bulletin board in the drama department of local colleges.

Provide a list of questions for your receptionist so that he/she can prescreen applicants over the telephone. Some possible questions to ask, include:

- What is your experience as a performer?
- In what area do you live?
- Do you have reliable transportation?
- Do you own sound or lighting equipment?
- Do you have a music library and of what does it consist?
- Do you have a videotape of recent performances?

If the applicant passes your prescreening guidelines, set up an interview. The applicant should fill out a detailed job questionnaire prior to meeting with you. When you meet face-to-face, pay close attention to the person's punctuality, appearance, and ability to conduct her/himself in a confident, articulate, and professional manner.

If you think you will want to hire this person, check out their references. If they pass with "flying colors," hold a second interview, which can include a live-performance demonstration.

Dancers

Follow the same guidelines for hiring dancers as you use for DJs with one notable exception; also consider holding an "Open Call." During the audition make note of the various dance styles a person is capable of performing including participation dances, country line dancing, Latin, swing, disco, partner, and freestyle solo.

Your company's dancers should ideally be versatile, outgoing, and fun, and have great people skills and a "contest winning" smile. They should be adept at pleasantly motivating party guests. With a solid group of experienced dancers, you can upsell clients on purchasing this "entertainment extra" and be sure you have plenty of work for the folks you have hired.

DJ Compensation

Fair compensation for DJs generally ranges from $15 to $100 an hour, and is usually based on some or all of the following factors:

- Experience and ability
- Length of employment (work with company)
- Specific request from a client
- Event type
- DJ booked job for company
- X number of consecutive "excellent" ratings on client evaluations
- Using his/her personal CD music library and/or equipment on the job

Tips

All of your top DJs will receive frequent high tips from clients if you add verbiage to your contract that states, "Gratuities given to your DJ Entertainer are made at the Client's sole discretion. 10 percent is customary for an excellent performance." This informs your customer about what is customary and standard for tipping in our industry. It will also motivate your DJs to be their best because they understand that their tip is riding on their performance at an event.

Figure 8.1 Denon and Doyle Disc Jockey Company. Courtesy *DJ Times*.

Motivation and Recognition

You can motivate your employees by working to maintain a pleasant office environment and first class reputation. Strive to be a strong but compassionate leader, and set a positive example for your staff.

Provide your beginning DJs with a comprehensive training program and manual, and all of your staff with the tools that are necessary to do their jobs effectively.

Routinely observe your DJs' performances. Be long on praise and short on criticism. If someone demonstrates initiative with a new idea that will benefit your company, reward them appropriately and give him/her credit in front of other staff. People greatly appreciate praise for a job well done. This type of leadership will make you generally liked and respected and will go a long way toward motivating your staff to continually give their best.

Set performance standards that apply to everyone equally in your company. All of your company's policies, procedures, and standards should be set forth in a manual. Consider implementing a probationary period for all new employees and existing ones who have violated company policy.

Performance

The most effective way to provide leadership that improves performance is to coach people using a problem-solving approach rather than a disciplinary or punishment-oriented approach. First, identify the performance problem, and clearly state why it did not meet your standards or expectations. Ask the employee if they realized that there is a problem. Listen carefully to what the employee has to say. If the answer is "yes," move on. If the answer is "no," you have got a bigger problem. Once the employee recognizes the problem and agrees on a specific action to correct it, you should follow up on the employee to see if the employee made the necessary adjustments. The key here is to stay positive but firm. Keep your conversation constructive and positive.

Be certain to stay away from language that speaks to personality or personal style. Ask the employee to give you her/his own solutions to the problem and then commit to implementing those solutions immediately. This method is far more effective than a management style that dictates solutions to subordinates.

Be empathetic but firm in the meeting. Go over your action plan with the employee and the timetable associated with it. Be clear about the consequences of noncompliance with the plan (such as disciplinary action) for future infractions. Document the conversation and what the employee agreed to do to solve the problem, then follow up by providing support and monitoring the improvement objectives and timetable.

By providing a positive work environment and treating people equally and with compassion and consistency, your employees will feel that they are part of a professional and well-respected team.

Figure 8.2 DJ icon Bernie Howard is geared up to provide Emergency Music Support with his "Jambulance."

Disciplinary Action and Termination

If a member of your staff is demonstrating performance problems, do not wait long to deal with the situation. Review the effect that the performance problem has on the company, projecting a positive attitude toward the employee that your only interest is to bring their performance back up to par. This step is especially important for employees who normally have good work habits or when there is a minor infraction.

If the problem persists you may need to terminate an employee. Unless the person has committed an act involving theft or violence, always handle terminations face-to-face. Once you tell an employee you are terminating his or her employment, he or she should be allowed to gather personal property (under your supervision) and then promptly leave your place of business. Arrange to have a third party present so that you have a witness in case any questions arise at a later date.

When, as a Manager you find it necessary take disciplinary action against an employee, you may likely be met with some form of hostile reaction. Maintain a calm, cool, and collected demeanor. When confronted, many people will plead ignorance and claim that they do not remember the situation you are bringing to their attention.

Begin by complimenting the employee on the positive things that they are doing and tell him or her that he or she is a valuable part of your team. Whenever a performance problem that requires disciplinary action is discussed, it should also be documented and placed in the employee's file. Be certain that you can provide specific dates and details when confronting someone. This will support the resolution of any dispute that may arise in the future.

No-Compete Contracts

Many mobile DJ companies ask their employees to sign no-compete contracts. They do this to ensure that the time and money they spend training a DJ will not become an investment in someone else's business.

If you want to be able to enforce your no-compete contract, you must have a formal company training manual and documentation of the training provided. Even so, if the employee is ever terminated or wishes to leave your company to work for him/herself or another mobile service, no-compete agreements are difficult to enforce because the laws are generally in favor of a person's right to be able to earn a living by working at their chosen vocation.

Creating an Employment Manual

Every company that has employees should also have an employment manual. There are software programs on the market that can provide you with a boilerplate manual. You need only fill in the specifics of your company's policies with these programs.

As an employer, it is wise to use an employment agreement. This should address all of the contractual obligations between the company and its employees. Be sure to confer with an attorney regarding the legal validity of your company's employment manual and agreement.

Treat Your Employees Well

The best ways to attract and retain employees is to TREAT THEM WELL! This is accomplished by compensating them fairly, showing appreciation and respect for the contribution they are making to your company, and providing opportunities for them to grow as professionals.

DJ Training Program

The manager of a multi-system mobile DJ service faces multiple operations and management challenges. It is critical to the success of any company that communication be clear and equal to all members of the organization. By developing a comprehensive training manual and program you will reduce the types of mistakes that are frequently made by rookies. Excellent resources to help you get started can be found on a variety of websites that cater to mobile DJs.

It is best to teach in a manner that builds new skills upon ones that have already been learned. There are many levels to mastering a given skill, which can only come through time and experience.

Effective teachers give real-life examples to present necessary information being presented in a form to which students can relate. They also demonstrate, by example, the physical tasks involved in what is being explained. Ample opportunity is given to students to ask questions and to practice, practice, practice—both in front of their classmates and on their own. Test students to ensure that they thoroughly understand what they have been taught. This should include written and performance tests.

It is helpful to divide training sessions into increments. For full-day training you might consider doing 1–1½ hours of lecture/classroom time, followed by a 15-minute break. Then 1½–2 hours of hands on training, followed by 1½ hours for lunch. After lunch begin with more classroom training, and end the day with hands-on instruction.

A comprehensive mobile disc jockey training program combines classroom learning, hands-on training, and an apprenticeship. The training program should cover everything a mobile DJ needs to know in order to operate as a professional. Classes should include:

- How to properly operate sound and lighting equipment.
- How to "read a crowd" in order to play the music to which people will dance.

- How to break the ice with an audience.
- How to teach popular dances.
- How to run contests and games.
- How to run a wedding reception.
- How to run a Bar/Bat Mitzvah reception.
- How to segue and beat-mix music.
- How to deal with difficult people.
- How to dress and act professionally.
- How to follow the company's expectations for its DJs.

Classroom lessons should progress in a logical order. They should cover topics such as musical programming, microphone technique, effective interaction with guests, and how to generate personal referrals. *The Mobile DJ Handbook* can be a useful tool in your training curriculum.

Before graduating, students usually serve an apprenticeship with working DJs. It is best for each student to work with at least three different disc jockeys to observe their various styles and to ask questions. It is also recommended that a student have the opportunity to observe different types of events, such as a wedding reception, anniversary party, prom, and Bar or Bat Mitzvah. After each event, quiz the student to determine what he or she learned. Obtain feedback from the DJ on the student's performance. This will help to determine if the student is ready to graduate and perform.

The last phase to a DJ training class is ordinarily the DJ audition test. This usually includes simulated performances of various events in front of the other members of the class who will act as the guests at an event. An audition test also generally includes a written test as well. If a DJ passes the test, award a diploma and, if a job opening is available, a position with your company. If a student does not pass, focus on the area(s) that need improvement, then give the test to the DJ again.

Office Relocation

Your accountant can help you determine when it is financially feasible for you to move out of your home office and into another business location. Leasing office space can be expensive. It is best if you wait until you can afford to make this move.

There are some important factors to consider when searching for the ideal space. These include cost, easy access and ample parking for your clients, sufficient working space for your business, and storage space for your equipment and music. You may not need a ground-level storefront (very expensive), but the eleventh floor of a high rise is probably not the ideal location for your disc jockey business either.

Consider taking over a small, existing building or sharing a small building with one or two other businesses. It may be wise to choose a location with a photographer, videographer, or some other business that can work in conjunction with your mobile DJ service.

Multiple Systems and Inventory

As your business grows and you add new systems, you will also acquire all of the technical and logistical problems that go along with multi-system management. It is strongly suggested that you hire a part-time engineer to maintain and repair your equipment.

If your electronic purchases rise to a certain level, some manufacturers will allow you to become an authorized dealer for them, allowing you to buy their product at lower cost. If this prospect is of interest to you, contact the equipment manufacturers from whom you buy and inquire about how their dealerships are granted.

Advertising Changes

If your company has multiple systems or operates from more than one location, you may want to consider purchasing a large Yellow Pages display advertisement. Radio and television advertising may also be viable considerations as part of your company's expanded marketing and advertising program.

Radio

Radio permits you to reach an audience when they are in the mood and at a time of day in which they are most likely to be receptive to your message. If you choose radio as an advertising vehicle, buy or barter for airtime only on stations most listened to by your clientele. The niche you choose in the marketplace will determine the general demographics of your clients.

Radio stations that play Contemporary Hit Radio, Urban Contemporary, or Top-40 formats sometimes do live broadcasts from clubs. They often need a great sound system to entertain the crowd. Perhaps you can work out a trade deal by providing the system to one of these stations in exchange for commercial airtime.

A well-produced commercial can include mood music. It can also include testimonial soundbites from satisfied customers. Consider having a company jingle produced for use in all of your radio advertising. Sell the benefits of your unique DJ service. Tell listeners about your experienced DJ entertainers, state-of-the-art equipment, and large all-CD music selection. If you offer party- or event-planning services, dancers, video and lighting effects, or Karaoke, mention this in the commercial as well. If your service is geared toward specific types of functions (weddings, Bar/Bat Mitzvahs, corporate events, etc.), mention these as your areas of specialty in the radio spots.

A 10-second commercial is useful to keep your company's name familiar to the audience. It can comfortably accommodate 20 to 25 words. A 30-second commercial permits some explanation of a sales idea and at least one repetition of your business name. It can accommodate 55 to 75 words.

Radio stations offer discounts for large media buys. Discounts often depend on how frequently the commercials run and the total amount of airtime purchased. Stations offer various package plans that allow for a combination of prime time and non-prime time commercials, at a given number of times per week, for a specific number of weeks. Such plans offer considerable savings in comparison to purchasing individual spots. Quoted prices are always negotiable! Offer to pay 15 to 20 percent less than the total cost you are quoted. You will have additional leverage if you offer to prepay for your commercials.

Although researching, negotiating, and purchasing commercial airtime yourself is an option, it is recommended that your hire a media planner/buyer to do this for your company. These folks are experts on this subject matter and can recommend a plan based on your goals and budget. Furthermore, because they buy so much media for all of their clients, they can almost always negotiate a significantly better rate for your company than you can independently.

Run your radio advertising prior to the busy seasons. Air your commercials in the beginning of November to promote both

Christmas and New Year's eve parties. Early April is a good time to begin promoting proms, graduations, and weddings, which generally run from May to September.

Television

Purchasing television commercials is much the same as purchasing radio commercials. The same negotiation rules apply. A media planner/buyer can advise you regarding the TV shows, stations, and time slots that will best allow you to reach your target audience.

Effective television advertising will pay for itself by increasing inquiries that you can turn into bookings. It is imperative for you to have adequate staffing and equipment to accept the additional business.

Many television stations will write, produce, shoot, and edit a commercial for you free of charge or at a low charge if you advertise with them. This can be beneficial if you are working on a tight budget. If you choose this option, be sure you have approval rights on the completed commercial in your contract, and be aware that the production quality may not be the best.

It is recommended to have your commercial videotaped by a professional production company. This will cost several thousand dollars, but the product will probably be worth the cost. Today's viewing audiences are sophisticated and expect to be visually "wowed." Using graphics, creative visual images, and music will gain their attention.

A word to the wise: Do not include anything in your television commercial that will date the production, or is of variable information. Once you have gone to the expense of shooting the spot, you may want to run it for several years. You can have the production company shoot your commercial as a "donut." This is a commercial with an unchanging open and close, and a changeable center or "hole." The hole can be filled with information that changes from season to season or year to year. It can also be filled with information geared toward certain markets or events.

Franchising and Multiple Locations

Once you have perfected running a mobile disc jockey service into a highly profitable operation, perhaps it is time to consider opening up additional locations or franchising your business. Obtain the assistance of an accountant and an attorney for either of these two scenarios. If you are considering franchising, hire an attorney with a specialty in this area.

You can advertise the franchise of your DJ operation through DJ publications and advertisements in major city newspapers. Predetermine the qualifications for purchasing a franchise and screen prospects over the phone. Set up personal meetings whenever possible. At the meeting, provide a verbal and written overview of your company. Explain the advantages of joining a franchise operation in comparison with starting a new company. Your presentation should explain exactly what your franchise system would provide.

Consider providing the following to franchisers:

- Instructions regarding the hiring and training of DJs and other employees and subcontractors
- Information pertaining to employer responsibilities
- A monthly newsletter containing industry updates and helpful tips
- An excellent business track record and name recognition
- An exclusive territory
- A customized bookkeeping and accounting system
- A complete marketing and advertising program that includes camera-ready materials
- Personal training at your location
- Access to sound and lighting equipment at discount prices
- A list of songs each DJ should carry or the music itself
- Telemarketing and in-person sales techniques
- Insurance discounts
- Cellular phone and pager discounts
- Unlimited telephone consultations with your main office
- A training manual that explains every aspect of operations
- A booking and scheduling system
- Copies of *The Mobile DJ Handbook*

You must take action now
that will move you towards
your goals. Develop a sense
of urgency in your life.

LES BROWN

9

Selling Your Services

*Promise only what you can deliver. If you can't provide a
service the customer wants, help them find an alternative.
Honesty. Integrity. Professionalism. Three words to live by
in running your business.*

KEITH ALAN, OWNER
Keith Alan Productions
Prospect, Connecticut

Pricing Competitively

The pricing structure under which mobile DJs operate varies greatly
and depends on a number of different factors. These include pricing
by the competition in your market, the type of occasion you are
booking, the time of year, the day or evening and location of the
event, and the "extras" added to your primary rate. $100 to $150
is the average hourly rate in the industry; however, hourly rates
can range from $75 an hour to $500 an hour or more, depending on
lighting effects, dancers, props, Karaoke, and so on.

Experience and reputation are two factors to consider when
determining your pricing structure. Those with a newly formed
business may want to consider charging less than the rate of an
established business with years of experience and a solid reputation.
As your reputation and experience grow, slowly increase your
rates.

It is wise to base your prices on a 3-hour minimum with
higher rates during peak periods. These include April, May, June,
September, October, and December. Other peak periods include
New Year's eve, Friday evening, and Saturday day and evening.

133

Weddings, Bar/Bat Mitzvahs, and some corporate events are typically booked at a higher rate than other types of affairs. It is standard to charge a higher rate for functions where the entertainer has the dual responsibilities of being both the DJ and the emcee.

Sometimes a good incentive for people to hire your DJ service is to offer them a discount. Consider offering discounts to:

- Nonprofit organizations.
- Customers willing to prepay for an event.
- Corporate clients who book your DJ services four or more times in a calendar year (apply the discount to the last job).

You can earn additional monies through playing overtime at events. On site, about a half-hour before the job is over, ask the client if they want you to play overtime. The overtime rate is typically 50 percent of the hourly rate you have quoted for each 30 minutes of overtime you play.

It is extremely important that you do not underprice your service. As an industry, mobile disc jockeys need to better understand their value and not be willing to bend too far on price just to get a job. If every DJ added just 10 percent to the cost of their existing services, the entire industry would benefit from this price increase. I am not referring to "price fixing"; I am referring to creating an industry standard where there is room for low-, middle-, and high-end packages. Always be worth your asking fee and give your customers even more than you have promised.

Getting Paid What You Are Worth

As professional mobile disc jockeys we are also music programmers, masters of ceremonies, audio and lighting technicians, and event planners. What is more, we have technical and motivational skills, the ability to "read" a crowd to ensure our music and actions are in line with their desires. We can teach participation dances and host contests and games. Do we warrant a premium fee? You bet we do!

Selling over the Phone

If your marketing and advertising program is working, your telephone should be ringing with inquiries. This is the time to sell your services. Here are some tips for successfully converting telephone inquires into bookings:

- Ensure that the employees responding to your customers' inquiries are thoroughly knowledgeable about your company's services and products.
- Ask the potential customer how they came to learn about your business and write down the answer so you can track your advertising and marketing to determine cost versus return.
- Ask the potential customer if they have any specific needs or concerns. Write down all pertinent information.
- Promptly answer all questions with accurate information in a clear, concise, and professional manner.
- Always describe your services and products before quoting prices.
- Ask for the sale! If the person is unwilling to commit, attempt to advance the inquiry to an in-person consultation. If the person is still unwilling to commit, offer to send information in the mail and do so promptly.
- If information is mailed, do a follow-up phone call one week from mailing.
- If a booking was made, promptly send out a contract.

Here is an example of a typical inquiry:

CUSTOMER: "I'm calling around for prices of DJs. How much do you charge?"

COMPANY: "That depends on several factors. If you'll tell me the date of your event I'll check on our availability."

CUSTOMER: "June 6."

COMPANY: "Thank you. While I check on that date I'd like to take a moment to tell you about our company and the services and products we offer so you can decide which package is best for you." (Give a brief, concise description of your services and products.) "Which one of these packages most fits in with what you had in mind?"

CUSTOMER: "The one with 4 hours of music and the basic light package."

COMPANY: "Great. I see that we do have that date available. The cost for the package you mentioned is $500 for 4 hours.

Again, that includes (describe package). We offer our customers easy payment through Visa, MasterCard, or Discover credit cards, or by check or cash. Which is most convenient for you?"

CUSTOMER: "Personal check."

COMPANY: "That's fine. Would you like to go ahead then and book the party?

CUSTOMER: "I think so but I'll need to check with my husband."

COMPANY: "That's fine but I do need to tell you that June is a very busy month for us and most other professional DJs because of weddings and proms. I can't promise I'll still have the date open when you call back.

CUSTOMER: "Oh, I see. Well, I guess I'd better go ahead and book your services now."

COMPANY: "Good idea. I'll send out an entertainment agreement for you in the mail today. Please sign it and send it back along with a deposit of $100 within 10 days. I can only guarantee your date up until that time. May I please have your name and street address?"

CUSTOMER: "Nancy Drewski, 407 Detective Lane, Hardy, Arkansas, 84747."

COMPANY: "And what are your day and evening telephone numbers, Nancy, in case we need to get in touch with you? May I also please have your E-mail address?"

CUSTOMER: "Work is 000-0000. Home is 000-0000. My E-mail address is nancyd@yahaol.com."

COMPANY: "Could I please have the name and phone number of the facility where the event is going to be held?"

CUSTOMER: "It's going to be at the Hardy VFW in Hardy. Their phone number is 000-0000."

COMPANY: "Last question. When would you like your DJ entertainer to begin and end?"

CUSTOMER: "I'd like entertainment from 7:00 P.M. to 11:00 P.M."

COMPANY: "Okay, Nancy. We're all set for now. Like I said, I'll get the paperwork right out in the mail to you. We appreciate your phone call, and if you have any questions, please feel free to call and ask for me. My name is Fred Flintrock."

CUSTOMER: "Thanks, Fred."

COMPANY: "Thank you, Nancy. Good-bye."

CUSTOMER: "Bye."

You will know if your phone style is effective by the number of calls you convert into sales. Make a log of all inquiries and include:

- Those you immediately converted into bookings.
- Those you converted into an in-person consultation.
- Those to whom you mailed information.
- Those who would commit to nothing and did not want additional information.

By keeping a record of this information, you will be able to better determine the effectiveness of your telephone sales technique. This information is important. It may give you some helpful insights into where you may need to make adjustments.

Here are some quick answers to common objections:

CUSTOMER: "I want to check out other DJ companies before I make a decision."

COMPANY: "I'll give you a 30-day money-back guarantee in writing that states if you are able to find another DJ service equal or better to ours, we'll refund your money."

CUSTOMER: "I don't want information. I only want to know your price."

COMPANY: "We have four packages: $400, $600, $800, and $1200. Which one would you like to hear about?"

CUSTOMER: "How can I be sure of your quality and reliability?"

COMPANY: "Here are three references of satisfied customers we've recently done parties for. And . . . (refer to moneyback guarantee)."

In-Person Consultations

- Call the day before the meeting to confirm the directions and meeting time.
- Dress in business attire.
- Show your company DVD/video first, then sit down to talk with the potential client.
- Discuss the information you obtained during the telephone inquiry.
- Show the potential customer your marketing materials and a reference list.
- At the conclusion of the meeting, ask for the booking and a deposit. When possible, immediately issue a contract and any additional information that has been requested, such as a music list.

Using DVD/Video as a Sales Tool

A polished Video, DVD, CD ROM, or DVD ROM of your company's disc jockeys in action can greatly enhance your ability to make a sale with clients who view it. The video needs to include all pertinent information about your service, the types of events that you do, and any special features or packages you offer. The video is especially important to close sales with wedding and Bar/Bat Mitzvah prospects.

You may consider having a different video or segment on a video for each type of function you perform. A senior class advisor will not necessarily be impressed by a wedding video, nor would a bride or groom be "wowed" by watching an outrageous prom video with lots of CHR and alternative music. Match the appropriate video or segment to the client and the occasion.

An enterprising way to get a wedding video made is to barter with a videographer at a wedding where you are booked. Ask that person to videotape the important parts of your performance. The client is already paying this person so the charge to you should be minimal.

A wedding video for a mobile DJ service should include wedding party introductions, motivating patter, the bouquet and garter toss, cake-cutting ceremony, your leading of participation dances, and footage of people dancing.

Strong consideration should be given to the creation of some form of company audio-visual tool. Be sure it is professionally produced and edited, and no longer than 15 minutes in length.

Handling Deposits and Payments

As a businessperson you must ensure that you are paid for your services. One way to do this is to always have clients sign a legally binding contract. This should be returned to you with a deposit within 10 days. The deposit should range from $100 to half of the total bill.

Nearly all customers will respect your professionalism if you explain in a friendly manner that you book on a first-come, first-served basis and that no date is reserved until the deposit and contract are received. It is essential to express both over the phone and in your contract when the balance is due. There will be greater clarity if this policy applies to all clients, including friends, relatives, and co-workers. It is best to have an event paid in full by the date of the performance. Some mobiles prefer to have the balance due on the date of the event prior to their start time.

In the event you receive an offer of a booking at the last minute because another company's DJ has canceled, arrange for the client to sign a contract as soon as possible and to pay by cash, cashier's check, or credit card. If this is not possible, make it the policy of your company to accept only cash or a cashier's check at the time of your arrival at the function.

See sample client contract in Chapter 11.

Generating Referrals

Many DJs find that after a couple of years, most of their business comes from referrals. Referrals will come from several sources and there are actions you can take to increase the number of referrals you will receive.

First and foremost, you and all of your DJ entertainers should consistently put on a stellar performance. Propose a reciprocal arrangement for referrals with videographers, photographers, banquet managers, caterers, bridal store and tuxedo shop owners or managers, and so on, where each of you receives a dollar amount or other incentive for every referral that leads to a booking.

Another, and perhaps the most profitable affiliation you can have, is with your past clients. A happy client who tells others about your service is your best advertisement. After every job, send the client a thank-you note with two business cards enclosed and a postage-paid evaluation card. The thank-you note can offer $25.00 off their next booking for every inquiry through them that leads to a booking. The evaluation card can ask brief questions about the company and the DJ's performance.

Send past clients a note reminding them of your offer, along with two business cards every six months to a year.

Tips for Maximizing Your Profit

- Keep your references, marketing materials, and music lists current.
- Have an ample supply of business cards and other printed materials that are "billboards" for your DJ service.

- Perform networking and telemarketing activities on a daily basis.
- Join clubs and organizations and attend their mixers.
- Acquire a great quality DVD/video and/or CD ROM/DVD ROM for your company.
- Conduct direct mail campaigns to specific target markets.
- Perform quarterly cost versus return analysis on your advertising and marketing campaigns.
- Read your local newspapers to generate ideas for wedding clients, corporate events, and other types of functions.
- Ask the people/vendors you spend money with to recommend your services. Offer incentives.
- Clean and maintain your equipment regularly.
- Add more "profit centers" to your business and become a one-stop entertainment shop.
- Listen to and read motivational and marketing CDs and books as well as the trade publications.
- Attend trade shows.
- Rent space at bridal shows.
- Subrent space in your existing location or seek out a successful noncompeting service that you can subrent from to reduce overhead.
- Value yourself and your services.
- Maintain balance between your work and professional life: mind, body, and spirit.
- Be charitable and choose a few organizations yearly for which you will DJ at no cost. When we give freely, our efforts come back to us multifold.
- Ask for the sale.
- Avoid burnout and have lots of fun.

A Sound Reinforcement Idea

There is a huge amount of money up for grabs in the sound reinforcement and equipment rental market because of the millions of dollars that corporations spend annually on events where they need this gear. They could be hiring your DJ service to meet their needs at a lower cost and with greater service than most equipment rental companies offer, so buck the trend!

For most mobiles, weekdays are "dead" times when their sound and lighting gear lies idle in a storage area. You can change this by adding sound reinforcement as an additional revenue stream for your business. Many folks just need a basic public address system. Providing this from your equipment arsenal is a "no-brainer," and you can probably offer a whole lot more.

Do direct mailings to corporations in your area. Contact banquet facilities and hotels, and send them a sound reinforcement rate sheet (along with your other marketing materials). Meet with the person in charge whenever possible.

Your three primary sales points are:

1. You (or a DJ from your company) will personally monitor the sound or lighting at the event.
2. Your sound reinforcement equipment rental rates are 10 to 20 percent lower than the companies in your area.
3. The quality of your gear is equal to or better than what others are offering in the marketplace.
4. Be sure to do the homework on your competitors' pricing so that the information you provide and rates you offer are accurate.

*Learning is not attained by
chance. It must be sought
for with ardor and
attended to with diligence.*

ABIGAIL ADAMS

10

Tech Talk

There are many DJs that can crossfade, beat mix, and scratch but only a distinct minority have technical mastery of their gear and mixing techniques. This takes self-education and practice. Don't gloss over the fundamentals or you will miss learning key elements, develop bad habits, and your overall skill set will be poor. New forms of technology are changing the ways DJs play and control music, bringing about a renaissance in the industry. DJs now have the tools to create, remix, and deconstruct music in previously unprecedented ways. With power comes responsibility. Now, more than ever, gaining a solid understanding of the technical basics is essential for individual growth and for the growth of the art form.

GERALD "WORLD WIDE" WEBB
(The world's first digital turntablist)
www.digitalscratch.com, founder
Party Time Entertainment,
Williamstown, New Jersey

Sound Reinforcement Terms

A–B Test

Evaluating comparative performance of two or more models of equipment such as amplifiers or speaker systems by listening and switching quickly from one to the other. It is possible to connect an A–B test to any two pieces of the same type of equipment from anywhere in the audio chain.

Absorption

The ability of a room to take up or absorb the acoustic energy radiated within it. There are many types of absorption because it can be frequency dependent. There are certain materials such as acoustical ceilings that may absorb more high frequencies than lows, or diaphragmatic absorptions (caused by loose wall panels or cavities behind the panels) that cause certain low frequencies to be absorbed.

AC Mains

110–120 volts alternating current (60 Hz) (what you plug your power cord into).

Acoustical

Sound or properties of sound; the acoustical response of a room has to do with the way that room responds to sound.

Active

A type of electronic circuitry that can increase the gain or amplitude of a signal: Active gain controls, active equalization, active direct boxes, active crossover.

Ampere

A unit of measurement of electrical current (I). One amp of current represents 6.2818×10 electrons flowing past a given point in 1 second and is equal to 1 coulomb.

Amplifier (ampl)

A device capable of increasing the gain (magnitude) or power level of a voltage or current that is varying with time (frequency), without distorting the waveform of the signal. The amplifier is, just as the word implies, a signal amplifier. The incoming signal from any program material source is far too weak to power a speaker system. The role of the amplifier is to take that weak signal and strengthen it to the necessary power level to operate the loudspeakers with minimal distortion.

Analog

A physical variable that remains similar to another variable insofar as the proportional relationships are the same over some specified range. The electrical signal produced by a microphone is an electrical analog of the acoustic sound that the microphone is reproducing. The continuous electrical signal that the microphone produces varies in voltage and frequency as a direct correlation to the nonelectrical acoustic information impressed on the transducer. The electrical signal is analogous to the acoustical sound that the microphone reproduces. For example, the voltage that the microphone produces is the electrical analog of the acoustic sound source.

Attenuation

The reduction in level of a signal.

Audio Chain

The order of sequence for connecting audio components, i.e., microphone, preamplifier (mixer), effects device, graphic equalizer, crossover, amplifier, and speaker.

Audio Range

20 Hz to 20,000 Hz. (20 cycles per second to 20,000 cycles per second). (This is the frequency response spectrum of human auditory perception.)

Aux Input

An auxiliary input that serves as a straight connection to a signal bus (for instance, monitor aux input is an aux input to the monitor bus).

Baffle

The panel on which the speaker is mounted within the speaker enclosure. The term derives from its original use in preventing or baffling the speaker's rear sound waves from interfering with its front waves.

Balanced Cable

A pair of wires surrounded by a braided shield.

Balanced Line

A transmission line consisting of two conductors plus a braided shield, capable of being operated so that the voltages of the two conductors are equal in magnitude (voltage) and opposite in polarity with respect to ground. A balanced line offers common mode rejection or cancellation by attenuation; signals are electromagnetically induced into the signal lines.

Bandpass

Refers to a type of filter that passes a certain band of frequencies uniformly and attenuates, or reduces, the level of frequencies below and above the specified bandpass.

Bandwidth

Response characteristic in which a definite band of frequencies, having a low frequency and high frequency limit, are transmitted or amplified uniformly.

Beat Mixing

By using the pitch controls on a CD player or turntable, you can match the beats of songs that have similar BPMs. There should be a smooth transition from one song to another without the audience ever noticing. While the first song is playing, listen to the second in your headphone. Vary the pitch control at the point you intend to start the second song until the beats are right on top of each other. Use the crossfader on your mixer to bring up the volume of the new song until it matches the volume of the one that is already playing. Then, start to bring down the volume of the original song. Beat mixing is an especially important skill for a club DJ.

Beats Per Minute (BPM)

To determine the number of beats per minute in a song, use a watch with a secondhand or a stopwatch. Count the number of upbeats and downbeats in the song for 15 seconds, then multiply that number by 4 to equal a minute.

Bi-Amp

Separation of the audio spectrum into two bands, i.e., high frequencies (high pass) and low frequencies (low pass) by means of an electronic crossover, using two separate amplifiers or channels of an amplifier. One amp or channel is used to amplify and project the high pass signals (high frequencies) from the high frequency component or horn of the speaker system and the other amp or channel amplifies the low pass signals (low frequencies) and projects them from the woofer or low frequency component of the speaker system. This results in increased headroom and dynamic range.

Bits

Bits is a digital term. The more bits there are, the better the sound. A 12- to 16-bit sampler is needed for quality sound.

Boost

A term used to indicate an increase in gain of a frequency, or band of frequencies, when equalizing an audio signal. (Opposite of "cut.")

Bridge Mode (mono)

Operating a stereo amplifier in mono via the bridge mode switch, which then makes channel-A output the positive power rail and channel-B output the negative power rail. Since the signal swings between A and B channels, the output of the amplifier is twice that of single channel operation.

Bridging

Connecting one electrical circuit in parallel with another. (Example: Paralleling power amplifier inputs.)

Bus

A conductor that serves as a common connector to several signal sources, most often associated with a mixer. A separate signal routing to a specified output.

Cardioid

A type of microphone having a heart-shaped pick-up pattern that picks up sound better from the front (on axis) than back (off axis).

Clipping

Amplifier overload causing a squaring off or undesirable change in the waveform, resulting in distortion or perceptible mutilation of audio signals.

Cluster

An array of loudspeakers or horns suspended above an audience to act as a single or point source of sound.

Comb Filter

When two combining sound waves have different amplitudes, phases, and frequencies, the resultant sound wave develops many nulls or spaces where the energy has cancelled. When viewed on a graphic recorder the resultant frequency response resembles a comb due to the nulls or notches of information that have cancelled.

Compression

Reduction of the effective gain of an amplifier at one level of signal with respect to the gain at a lower signal level.

Compressor/Limiter

A compressor/limiter controls the dynamic range of the music playing. The music's quietist and loudest sounds define the dynamic range. This component ensures that the loudest parts of your music sources do not exceed a predetermined sound level. This prevents feedback and blown speakers.

Conductor

A wire, cable, or other material (metal, liquid, gas, or plasma) suitable for carrying electric current.

Continuous Power

This power rating represents the most conservative statement of the capability of an amplifier. It is also called rms power. It denotes the amount of power an amplifier can deliver when amplifying a constant steady tone. It is usually measured at a signal frequency of 1000 Hz for a specific distortion. ($W = V^2/R$ or power in watts equals the voltage squared divided by the resistance of the load.)

Continuous Program Material

A signal, such as speech or music, that continuously changes in both frequency and voltage (time and amplitude).

Crossfading

This is the process of starting one song before another has ended. Fade the volume up slowly. As the first instrumental beat hits, you should be at full volume and immediately crossfade out of the previous song.

Crossover (x-over)

An electronic device that is used to separate an audio signal into two or more bands of frequencies or component signals above and below a certain frequency, said to be the crossover frequency or crossover point. Crossovers can be active or passive.

Crossover: Active

Electronic or active crossovers do not have the problem of excess power because only the power needed by the driver must be generated by the amplifier. An active crossover is employed when bi-amping a system. The active crossover separates the audio spectrum (full range) into bands of frequencies above (high pass) and below (low pass) a certain frequency (crossover point). The low pass is rolled off (attenuated) at a rate of X-many dB per octave above the crossover frequency. The high pass is rolled off (attenuated) below a certain crossover frequency at a rate of X-many dB per octave. The high pass and low pass outputs of the electronic (active) crossover are connected to the inputs of two separate power amplifiers whose respective outputs are used to drive the high end (horns) or low end (woofers) of a sound system.

Crossover: Passive

A passive crossover is built into most speaker cabinets in order to separate bands of frequencies from the full range speaker level signal that are produced by the power amplifier, and then routing those bands of frequencies to the proper speaker or driver. Most common speaker crossovers use iron in the inductors to decrease their size. This can be a source of distortion due to the nonlinearities in the coil from core saturation. The power going to the high frequency drivers must be attenuated due to the increase in efficiency of a high frequency driver as compared to a bass driver. This power has to go somewhere and it is usually converted into heat with the use of resistors.

Cueing

Cueing is setting up your music to begin at the precise moment you press the start button. Turntables require a one-eighth to one-quarter turn counterclockwise; cassettes one-quarter clockwise turn on the left hole; CD and MD players start immediately.

Current

(I) The rate of flow (measured in amperes) of electricity in a conductor or circuit. The amount of current that flows is determined by the voltage or electrical pressure applied and the conductivity of the substance or material (which also determines the resistance or opposition to current flow).

Cycle or Hertz

A unit of motion referenced to a time period of 1 second. The frequency of a vibration or oscillation in units per second. One hundred hertz (100 Hz) or 100 cycles per second (100 cps) refers to the number of times per second that a string is vibrated or an amplifier swings between its positive and negative supply voltage.

Damping Factor

The ratio of the speaker impedance to the amplifier's internal output impedance. Damping factor is a measure of how well an amplifier can actually control the movement of a speaker cone or diaphragm by preventing it from moving farther than intended. Dividing speaker impedance by the amplifier's internal output impedance derives damping factor. The internal output impedance of any amplifier is determined by the transconductivity (internal resistance) of the output devices. The speaker interprets anything connected in the speaker line (including the speaker cable itself or a crossover) as an increase in the output impedance of the amplifier, thus lowering the effective damping factor. Because any speaker is a mechanical device, it will have its own resonant frequencies, which will cause the cone to continue in motion after a musical signal has stopped. (See *transient distortion*.) An amplifier with a high damping factor will damp out these speaker tendencies.

dB (decibel)

A unit for describing the ratio of two voltages, currents, or powers. The decibel is based on a logarithmic scale. When measuring differences in sound pressure level (spl), the amount of change in sound pressure level perceivable is directly proportional to the amount of stimulus (the more sound present, the greater the change must be to be perceived).

0 dB

In the measurement of spl or sound pressure level, 0 dB is referenced to the threshold of hearing or auditory perception of a tone of 1000 cycles (Hz) per second (1 kHz). Zero decibel (0 dB) must always be referenced to some base of measurement. In gain functions 0 dB is unity gain (1).

3 dB

The amount of spl gained by doubling the power to a speaker. The amount gained by doubling the number of speakers.

±3 dB

Plus or minus 3 dB is a measurement of frequency response that exhibits no more than +3 dB and no less than −3 dB below a given reference. It is actually a 6 dB window. The response of 60 Hz to 14 kHz ±3 dB means that within the bandwidth of 60 cps (cycles per second) to 14,000 cps, no frequency is +3 dB more nor −3 dB less than a specified reference frequency.

Diaphragm

A thin flexible sheet that can be moved by sound waves, as in a microphone; or can produce sound waves when moved, as in a loudspeaker or compression driver.

Digital

Refers to the processing of audio signals as having discrete values as opposed to a continuous analog audio signal. In digital audio the continuous analog signal is converted to an encoded discrete value or digital word.

Direct Field

The sound that emanates directly from a sound source or loud-speaker.

Directivity

Area of coverage of a speaker or microphone.

Distortion

Any undesired change in the waveform of an electrical signal passing through a circuit or transducer. Any distortion can be defined as deviation from the original sound, the discrepancy between what the amplifier should do and what it actually does. All distortion is undesirable. Distortion occurs when the amplifier alters the original sound in the process of amplification so that what comes out of an amplifier is no longer a true replica of what went in. Performers, however, will sometimes desire the application of electronically induced distortion for extra musical effect in the production of their "sound." The undesirability of inherent distortion is associated with high fidelity. It should not be confused with the desirability of distortion as it is expected to be produced through circuitry. When reproducing sound, distortion is unwanted.

Driver

The motor structure portion of a horn-loaded loudspeaker system. It converts electrical energy into acoustical energy and feeds that acoustical energy into the entry of a horn throat or the narrow end of the horn. Most often used when referring to a high frequency compression driver, called a driver for short. The definition also includes the loudspeaker in a horn-loaded woofer or mid-bass horn.

Dynamic Range

In a musical instrument, the dynamic range is the difference in decibels between the loudest and softest level of notes that can be played on that instrument. In electronic equipment, dynamic range is the difference in decibels between the highest (overload level) and lowest (minimum acceptable) level compatible with that signal system or transducer.

Echo

A delay in sound of more than 50 ms (milliseconds) resulting in a distinct repeat or number of repeats of the original sound.

Efficiency

The ratio, usually expressed as a percentage, of the useful power output to the power input of a device.

Electronically Balanced Input

A differentially balanced amplifier; an amplifier whose output is proportional to the difference between the voltages applied to its two inputs. It offers common mode rejection or attenuation of interference signal that was introduced electromagnetically in the signal carrying conductors.

Enclosure

An acoustically designed housing or structure for a speaker.

Equalization

The act of obtaining a desired overall frequency response through the implementation of graphic equalizers or tone controls. The word "equalization" implies balance, so when you "equalize," you balance the audio spectrum.

Exponential Horn

A speaker designed to reproduce high frequency sound. An exponential horn has a flare rate that increases with the square of the distance measured from the entry to the horn throat.

Feedback

Electronic feedback: The return of a portion of the output of a circuit to its input. Acoustic feedback: The squeal of a sound system caused by the regeneration of a signal from the output of a sound system into a microphone input.

Filter

An electrical or electronic device that permits certain frequencies to pass while obstructing others, such as a crossover filter used with loudspeakers.

Frequency

The number of vibrations or oscillations in units per second, which is measured in cycles or hertz per seconds. It is the rate of repetition in cycles per second (Hz) of musical pitch as well as of electrical signals. For example, the number of waves per second that a vibrating device, such as a piano or violin string, moves back and forth to produce a musical tone.

Frequency Response

A measure of the effectiveness with which a circuit, device, or system transmits the different frequencies applied. The way in which an electronic device (microphone, amp, or speaker) responds to varying frequency signals. This is a measurement of how well an amplifier reproduces and amplifies a specified audible range with equal amplitude or intensity, for example, between 30 to 16,000 Hz.

Full Range

The entire audio spectrum, 20 Hz to 20 kHz.

Gain

An increase in strength or amplitude (voltage) in a signal. The increase in signal power that is produced by an amplifier; usually given as the ratio of output to input voltage, current, or power expressed in decibels.

Ground

A heavy cable connected to earth via a metal copper stake to "ground" electrical equipment. In the United States, the third wire in our electrical system is the ground wire, which, when connected to the chassis of electrical equipment, provides protection against hazardous electrical shock.

Ground Loop

Hum caused by return currents or magnetic fields from relatively high-powered circuits, or components that generate unwanted, noisy signals in the common return of relatively low-level signal circuits. A potentially detrimental loop formed when two or more points in an audio system that are nominally at ground potential are connected by a conducting path.

Headroom

The difference between the average operating power level of an amplifier circuit and the point at which clipping or severe distortion occurs.

Hearing

The human hearing system is well designed. It has a dynamic range of over 120 dB. Contemporary digital recording techniques can only achieve a dynamic range of about 90 dB. The typical threshold of pain is around 140 dB, with discomfort starting around a sound level of 118 dB.

Hertz (Hz)

A unit of measurement, previously referred to as cycles per second (cps), used to indicate the frequency of sound or electrical wave. A unit of motion referenced to a time period of one second. The frequency of a vibration or oscillation in units per second.

High Z or High Impedance

Any resistance to ac voltage or current generally greater than 2000 Ω (ohm).

Hum

An electrical disturbance that can occur in sound equipment due to the frequency of the power distribution system or any number of its harmonics. Our power line frequency in the United States is 60 Hz. Hum can occur at 60 Hz, 120 Hz, 180 Hz, 240 Hz, and so on.

Impedance

The total opposition to alternating current flow presented by a circuit. It is the resistance to the flow of alternating current in an electrical circuit, generally categorized as either "high" or "low," but always expressed in ohms. Commonly used to rate electrical input and output characteristics of components so that proper "match" can be made when interconnecting two or more devices, such as a microphone, loudspeaker, or amplifier.

Input Overload

Distortion produced by too strong a signal from the output of a microphone or other signal source, such as a keyboard, connected to the input of a preamplifier.

Insertion Loss

A loss in gain of a system after a component has been added or inserted in the system. Insertion loss is loss of headroom.

Integrated

A type of design in which two or more basic components or functions are combined physically as well as electrically, usually on one chassis as opposed to a separate mixer and power amplifier.

Jack

A receptacle on a receiver, tape recorder, amplifier, or other component into which a mating connector can be plugged.

LED

Light-emitting diode, a semiconductor device that produces visible light when voltage of a certain polarity and potential is applied to it.

Limiter

An electronic circuit used to prevent the amplitude of an electronic waveform from exceeding a specified preset level while maintaining the shape of the waveform at amplitudes less than the preset level.

Linear

Having an output that varies in direct proportion to the input.

Line Out

An output connection found on mixers and preamps that provide an output level sufficient to drive the input of a power amplifier.

Load Impedance

The actual impedance of the load or speaker to which a power amplifier is connected, thus driving a load of a certain impedance.

Loops and Sampling

Sampling is when you have a mixer with the capability of recording and playing back a few moments of music whenever you press the sample button. Whatever sound you capture in the memory will continuously repeat if you keep playing back the contents of the memory. This is described as a "loop."

Loudness Control

A volume control with special circuitry added to compensate for the normal decreased hearing ability of the human ear at the extreme ends of the audio range when listening to lower sound levels. A typical loudness control boosts the bass frequencies and, to a lesser extent, the high frequencies. Sometimes this control is called "contour."

Loudspeaker Efficiency

The ratio of signal output to signal input used to state the power needed to drive a loudspeaker. Expressed in percentage. An example: power output = 2 W (watts); power input = 10 W; ratio = 2/10 or 20% efficiency. Efficiency can vary from 2 percent to as high as 25 percent.

Low Pass

All of the frequencies below a given crossover frequency.

Master

Main level or gain control for a bus or mix.

Microphone

A microphone is a transducer that changes acoustical energy (sound) into electrical energy.

Mixer

A sound-reinforcement device that has two or more signal inputs and a common signal output. Used to combine separate audio signals linearly in desired proportions to produce an output audio signal.

Mode

Another word for room resonance. When sound energy is restricted by boundaries, such as walls, floor, and ceiling, waves develop at certain frequencies or wavelengths, which are integers of the distance between the room boundaries. Room modes, or resonance, cause standing waves because once the wave is generated it stands there. The positive pressure peaks (antinodes) and negative pressure troughs (nodes) stay stationary within the boundaries.

Monitor

A loudspeaker or system of loudspeakers that permit the performer to evaluate or monitor his sound alone, or in conjunction with other sounds, and is mixed to the listener's preference by means of a separate monitor or reference mix.

Mono

Monophonic Sound. Sound produced by a system in which one or more microphones feed a single signal processing amplifier whose output is coupled to one or more loudspeakers.

Music Power

This is a power rating generally applied to high fidelity amplifiers for tones of short duration. It takes into account the fact that most amplifiers can produce a greater amount of power in short bursts than they can continuously. The rationale is that music is made up of such bursts rather than sustained single frequencies. It is higher than continuous power ratings for the same amplifiers. It is measured at a signal frequency of 1000 Hz for a specified distortion.

Noise

Any extraneous sound or signal that intrudes into the original as a result of environmental noise, distortion, hum, or defective equipment.

Ohm

The unit of electrical resistance, equal to the resistance through which a current of one ampere will flow when there is a potential difference of one volt across it. Ohm is the unit of measure used to express opposition to current flow. Every wire or part through which electricity passes has some resistance.

Omnidirectional

Applied to microphones to refer to uniform pick up of sound from all directions.

Output Impedance

The internal output impedance of an amplifier presented by the amplifier to the load. (Output impedance is often used incorrectly instead of load impedance.)

Overlaying

Playing two different songs with the same beat at the same time. Start one song and then cue up the other. Listen in your headphone and use the pitch control to make sure the beats are precisely matched. When they are, bring up the volume on the second song. You can also turn this effect into backbeating by having one of the two songs lag behind the other by two to four beats. The effect should be that you hear one beat and two vocalists—a primary and an echo.

Parallel

An electric circuit in which the elements or components are connected between two points with one of the two ends of each component connected to each point.

Passive

An electronic circuit composed of passive elements, such as resistors, inductors, or capacitors. Does not contain any active elements, such as vacuum tubes or transistors, which generally result in a signal loss.

Peak

The maximum instantaneous value of a signal amplitude.

Peak Power

Peak power is a manufacturers' marketing ploy and has no bearing on the actual performance of a product. Usually peak power works out to be twice continuous power. Some of these same manufacturers have come up with yet another power term referred to as "Instantaneous peak power," which is a further inflated and equally meaningless specification. Amplifier power should be judged on an equal basis when comparing one amplifier with another. Disregard ad copy such as that just described.

Phase

Phase is the time interval between two related events. Two signals are in phase when they reproduce the same sound or signal simultaneously; they are out of phase to the extent that one leads or lags behind the other in time. A signal is said to be in phase with another when the voltage and current amplitudes begin at the same time and move in the same direction.

Phasing

Phasing is accomplished by playing two copies of the same song simultaneously. Start one song and then cue up the other. Listen in your headphone and use the pitch control on your CD player or turntable to ensure that both songs are precisely matched. When they are, bring up the volume on the second song.

Pitch

When a sound source vibrates at a certain rate, it is also causing the same vibration in the eardrum. The pitch (or tone) is what the ears hear when anything generates a vibration.

Polarity

The quality of having opposite poles. In electromagnetic mechanical systems, some form of potential is referenced to one of two poles with different (usually opposite) characteristics, such as one that has opposite charges or electrical potentials, or opposite magnetic poles.

Power

Electrical energy, measured in watts, such as the current from an amplifier used to drive a loudspeaker. Power in watts: $W = V^2/R$.

Power Amplifier

The final active stage of the audio chain, designed to deliver maximum power to the load or speaker impedance for a given percent of distortion.

Preamp (preamplifier)

An amplifier whose primary function is boosting or amplifying the output of a low level audio-frequency source (such as a microphone), so that the signal may be further processed without appreciable degradation of the signal-to-noise ratio of the system. An amplifier that increases electrical signals from a microphone or other instrument to a level usable by a power amplifier. Preamp levels are approximately .1 V.

Resistance

Opposition to the flow of electrical current. Measured in ohms.

Resistor

An electronic component designed to have a definite amount of resistance; used in circuits to limit current flow or to provide a voltage drop.

Resonance

A tendency of mechanical parts, loudspeaker cones, enclosure panels, or electrical circuits to vibrate at or emphasize one particular frequency, every time that frequency, or one near it, occurs.

Response

The range of frequencies to which an amplifier or speaker will respond, and the relative amplitude or intensity with which these frequencies are reproduced.

Reverb (reverberation, acoustical)

The prolongation of sound at a given point after direct sound from the source has ceased, due to such causes as reflection from physical boundaries. Electromechanical reverb: An electromechanical device usually employing springs that randomly reflect the greatest amount of sound as possible, therefore simulating natural reverberation. Digital reverb: An electronic reverberation-effects processor that uses digital electronics to introduce the multiple delay paths.

RIAA

Stands for Recording Institute Association of America. A type of preamplifier used for turntables. It is necessary to use an RIAA preamp when using a magnetic cartridge.

rms (root means square value)

The square root of the time average of the square of a quantity; for a periodic quantity the average is taken over one complete cycle. rms voltage is .707 times the peak voltage of a sine wave.

Scratching

Slip-cueing a record back and forth with the volume increased is called scratching. A scratching sound effect is sometimes contained in a mixer with sampling capability. It can also be achieved with a turntable, using a special scratching needle.

Segue

A segue (pronounced seg-way) is a continuous flow of music in which one song overlaps another. A segue should be as tight as possible.

Sensitivity

The minimum input signal required to produce a specified level of output. In an amplifier the input sensitivity is the amount of voltage at the input necessary to drive the amplifier to its rated power output. Loudspeaker sensitivity is the power level necessary to produce a stated SPL at a given distance from the loudspeaker, usually rated at 1 W per meter.

Signal-to-Noise Ratio (S/N)

The ratio of the amplitude (or level of a desired signal) at any point to the amplitude or level of noise at the same point.

Slip Cueing

Slip cueing is accomplished with professional quality turntables. It is the same as cueing, except the turntable never stops. Find the point in a song where you want to begin, then move the record counterclockwise one-eighth to one-quarter of a turn. Hold the record motionless with the volume turned down. A moment before you want the music to start, increase the volume and release the record.

Snake

A multiconductor shielded input cable used when it is necessary to locate a mixer a long distance from the stage and the microphones.

Sound

A pressure wave motion in an elastic medium (air) producing an auditory sensation in the ear by the change of pressure at the ear. In short, sound waves are produced by a vibrating body in air.

Spectrum

Refers to a particular band of frequencies. The normal acoustic sound spectrum is the range of human auditory perception (20 Hz to 20,000 Hz). There is also a subsonic spectrum (considered to be below about 40 Hz) and an ultrasonic audio spectrum (above 20,000 Hz).

Splitter

A box into which one microphone or signal is connected that has two or more individual outputs available for that signal. Used when a separate monitor mix is required.

Stereo

In a sound reproducing system, stereo refers to the use of two separate signal processing channels driving two separate power amplifiers. These in turn power two separate speaker systems. However, most times in sound reinforcement, a stereo mixer is employed to drive a mono (single channel) system in order to have (submixes) separate instrument versus vocal mixes of the program.

THD (Total Harmonic Distortion)

When a single frequency of specified level is applied to the input of a system, the ratio (of the voltage of the fundamental frequency to the voltage of all harmonics) observed at the output of the system because of the nonlinearities of the system; THD is expressed in percent.

Transducer (X-DCR)

Any device or element that converts an input signal into an output signal of a different form. A transducer changes energy from one form to another. A microphone is a transducer that changes acoustical energy (sound) into electrical energy (voltage). A loudspeaker is a transducer that changes electrical energy into mechanical energy, producing sound or acoustical energy.

Transformer (X-FMR)

An electrical component consisting of multiturned coils of wire placed in a common magnetic field (medium) that will transfer electrical energy from one electrical circuit to the next. A transformer will only pass alternating currents (ac) and will not pass direct current (dc). By adjusting turn ratios a step up or down, condition of voltage can be achieved.

Transformer Balanced (X-FMR BAL)

An input or output that is coupled using a transformer in a balanced configuration for operation. The voltages of the two conductors at any transverse plane are equal in voltage and opposite in polarity with respect to ground. A transformer-balanced input or output will offer common-mode rejection, which means any common-mode interference signal will not pass through the transformer because it will be cancelled out.

Transient Distortion

Transient distortion interferes with the amplifier's ability to accurately follow abrupt changes in volume, such as the sudden burst of sound when an instrument is first played. Minimum transient distortion is vital to clean and crisp overall sound.

Transient Response

The ability of an amplifier or loudspeaker to accurately follow abrupt changes, such as the sudden burst of sound generated by an instrument. Good transient response is vital to clear or crisp overall sound.

Tri-Amp

Separating the audio spectrum into three bands, i.e., high frequency, mid-band frequency, and low frequency by means of an electronic crossover and three separate power amplifiers to amplify the three outputs of the crossover (high pass, mid pass, low pass outputs) that drive three separate components of a speaker system. This results in increased headroom and dynamic range.

Unbalance Cable or Line

A single conductor cable with a surrounding shield that connects to ground. Such a system is called unbalanced because it cannot be balanced or offer common mode rejection.

Unbalanced Input

An input in which one of the two terminals is at ground potential or connected to the chassis ground.

Voltage

Voltage is a measurement of electrical pressure or the potential to do work. Voltage is sometimes called EMF or electromotive force. The familiar 120 V at a wall socket is an example of available electrical pressure. If the prefix "m" is used (as in mV), it stands for millivolts or thousandths of volts. Microvolts, abbreviated μV, are millionths of volts.

Volts (voltage)

Potential difference or electromotive force (EMF).

Volume

The intensity or loudness of sound.

Vu Meter

A meter that indicates the audio frequency power level or volume units of a complex electronic waveform.

Watt

A unit of the measure of power. The electrical wattage of an amplifier describes the power it can develop to drive a speaker. The greater the voltage capability, the higher the wattage. Amplifier wattage requirements are greatly dependent upon the speakers that will be used, the size of the listening room, and average loudness that will be played through the speakers. $W = V^2/R$.

Woofer

A low-frequency speaker specialized for bass or low-frequency reproduction.

XLR

A connector (sometimes called a cannon connector) used in interfacing audio components. The connector on a low impedance microphone is an XLR connector.

Many of the Sound Reinforcement Terms used here were provided courtesy of the Peavey Electronics Corporation Web site at www.peavey.com.

*Never say "oops" in the
operating room.*

DR. LEO TROY

11

How's Your Form?

Don't reinvent the wheel by attempting to create your own contracts. Instead, use the examples here or on the Internet to ensure that all of your bases are covered. However, be certain to first have an attorney review them for your specific business and marketplace to check for any loopholes that may apply. This will give you peace of mind and should be part of your success Formula.

JEFF STILES, DJ ENTERTAINER
Ultimate Entertainment
Dickeyville, Wisconsin
Writer, *DJ Times* magazine

Here are two words about contracts and forms: Use them! You are a professional, and therefore it is critically important to ensure that you and your DJ business are legally covered in all relevant matters. Be certain to examine each component of your contracts and ask, "Is this a reasonable requirement?" If you find yourself in a position where you would hesitate to sign such an agreement, then your clients and employees may likely feel the same way.

Just because you take a reservation over the phone, this does not guarantee that you will get a signed contract back with a deposit. There will always be a certain percentage of contracts that you send out that will not be returned. This is why it is important to include a contract due date.

After receiving a signed contract, be sure to call the event facility and inquire about your needed power requirements, load-in

EVENT QUESTIONNAIRE

CLIENT: *IT IS <u>VERY</u> IMPORTANT THAT YOU MAIL/FAX THIS QUESTIONNAIRE BACK TO US*
AS SOON AS POSSIBLE, BUT <u>NO LESS THAN 2 WEEKS BEFORE YOUR EVENT</u>.
IT WILL ALLOW YOUR DJ TO KNOW EXACTLY WHAT YOU WANT!

Client Name:_____   Event Date:_____

We have found that the best way to get the crowd involved at an event is to play music for the <u>DANCING MAJORITY</u>! Your DJ's song selections are based on advance and on-the-spot requests, as well as dance floor reaction. He/she will play as many of the most danceable requests as time permits.

1) What are the event's highlights and approximate timing? (Ex. 7 – 8 pm cocktails, 8 – 9:30 pm dinner, 11 - 11:30 pm raffle, 9:30 pm – 1 am dancing, etc.). Is there a theme? Is there something special about the crowd that your DJ should know?

_____
_____
_____
_____
_____

2) What are the "must play" songs such as crowd favorites, participation dances, theme songs, etc.?

_____
_____
_____
_____
_____

(PLEASE WRITE ADDITIONAL REQUESTS ON THE BACK OF THIS QUESTIONNAIRE OR ON A SEPARATE SHEET)

3) <u>On a scale of 1 to 10</u> (10 = most), rate each category in terms of their likely appeal to the DANCERS in your crowd.

_____ Current Dance Hits   _____ Oldies/Motown
_____ Disco/Dance Classics   _____ Reggae
_____ Top 40   _____ Alternative/Modern Rock
_____ Classic Rock   _____ Country
_____ Ethnic (CIRCLE: Irish, Polish, Italian, Latin, Jewish)   _____ Big Band/Swing
Other _____

4) <u>On a scale of 1 to 10</u> (10 = most), how much ENERGY and CROWD INTERACTION would you like from your DJ? _____

5) Proper DJ attire? Formal ___   Semi-Formal ___   Dressy Casual ___   Casual ___.

6) Will a sit-down meal or food from the buffet be provided for your DJ Entertainer? Yes ___   No ___.

Figure 11.1 Event questionnaire.

information, and facility opening information for the evening/day of your event.

Listings of the standard forms used by mobile disc jockeys can be found at www.themobiledjhandbook.com, www.djzone.net, and www.prodj.com. These should be used as templates to create your own forms. Their lists include contracts, wedding information sheets, comment sheets, Bat/Bar Mitzvah sheets, party planning sheets, event evaluation forms, and letters.

Here are a few sample contracts to consider using:

Sample Client Contract

ABC MOBILE ENTERTAINMENT

Street Address, City, State, Zip Code
Telephone Number, Cell Phone Number, Pager Number
E-mail address, Web site address

ENTERTAINMENT AGREEMENT

DATE MAILED:

THE PARTIES: This Agreement is for entertainment services for the event described below, between the undersigned purchaser of entertainment (Client) and ABC Mobile Entertainment (DJ). DJ agrees to furnish services to the following specifications and Client:

ENTERTAINMENT/OPTIONS:

CLIENT NAME:

COMPANY/ORGANIZATION:

ADDRESS:

DAY/EVE. TEL. NOS.:

TYPE OF EVENT: DAY/DATE:

FACILITY NAME/ROOM: FACILITY TEL. NO.:

FACILITY ADDRESS:

EVENT START TIME: DJ START TIME:

EVENT STOP TIME: DJ STOP TIME:

NOTES:

APPROX. NO. OF GUESTS/AGE RANGE:

TOTAL FEE: $
(A 5% discount may be applied if the total fee is paid in full at the time of signing this Agreement).

DEPOSIT DUE/DATE:
(A tentative reservation is being held for Client's event until this date. Entertainment Agreements received after this date are subject to availability.)

BALANCE DUE/DATE:

OVERTIME: $00.00/HALF HR. When feasible, Client requests for extended playing time during event will be accommodated. Payment is due at time of request, and may be made with check or cash.

GRATUITIES: Gratuities given to your DJ Entertainer are made at the Client's sole discretion. 10% is customary for an excellent performance.

PAYMENT: Payment may be made with personal/company/ cashier's check, or cash. Please **MAKE ALL CHECKS PAYABLE TO ABC MOBILE ENTERTAINMENT**. (There will be a $20.00 fee charged for all returned checks.)

TO CONFIRM THIS AGREEMENT: (1) SIGN ONE OF THE TWO AGREEMENTS (must be 18 yrs. or older). RETURN ONE SIGNED COPY TO DJ. Retain the second copy for your records. **(2)** THE DEPOSIT MUST BE RECEIVED BY THE DUE DATE TO GUARANTEE SERVICES FOR YOUR EVENT.

CANCELLATION: This Agreement cannot be cancelled or modified except in writing by either the Client or DJ. If Client initiates cancellation less than 60 days prior to the event, then only the deposit will be forfeited. If Client initiates cancellation less than 30 days prior to the event, then Client is responsible for the total fee OR may forfeit deposit only by signing a new Entertainment Agreement with DJ within 14 days of cancellation for a substitute engagement performed in the following 6 months. Rescheduling for events canceled due to inclement weather shall be accommodated without penalty whenever possible. Rescheduled events are subject to availability.

PROVISIONS: The Client shall ensure that: **(1)** performance facility provides DJ with a sturdy covered table approximately 2 × 8 feet in an area within 25 feet of performance area; **(2)** Table shall be within 25 feet of a reliable 15-amphere circuit (3-prong grounded) for audio and/or a reliable 20-amphere circuit (3-prong grounded) for lighting effects; **(3)** facility is open at least one hour prior to scheduled start time; **(4)** facility meets all federal and state safety regulations and has all appropriate music licenses and performance permits where applicable; **(5)** reasonable steps will be taken to protect ABC Mobile's equipment, personnel, and music, and crowd control will be provided if warranted; **(6)** for outdoor performances, shelter is provided that completely covers and protects ABC Mobile's equipment from adverse weather conditions.

Client accepts full responsibility and is liable for any damages, injuries, or delays that occur as a result of failure to comply with these provisions. In the event of circumstances deemed by DJ to present a real or implied threat of injury or harm to DJ, equipment, or recordings, then DJ reserves the right to cease performance until such time as Client resolves the threatening situation. DJ further reserves the right to deny any guest access to recordings or equipment. In the unlikely event the DJ's performance is delayed, liability is limited to providing Client with performance time equal to time lacking. DJ holds all appropriate insurances for its equipment and personnel.

PLEASE SIGN BELOW AND RETURN YOUR SIGNED COPY ALONG WITH THE DEPOSIT AMOUNT NOTED ABOVE. This Entertainment Agreement supercedes all others for the aforementioned event date and is subject to the laws of the State of (?).

The Client signs below on behalf their company/organization (where applicable).

Client Name: _____

                   Print

Client Signature: _____   Date: _____

On behalf of ABC Mobile Entertainment:

_____   **Date:** _____

            John Doe

How did you hear about ABC Mobile Entertainment?

_____

You may need to adapt this basic wedding party introduction to accommodate special situations or client requests.

Sample Wedding Party Introduction Sheet

ABC MOBILE ENTERTAINMENT

<u>WEDDING PARTY INTRODUCTION SHEET</u>

PLEASE WRITE NAMES IN THE EXACT MANNER AND ORDER
YOU WOULD LIKE THE WEDDING PARTY ANNOUNCED.
WRITE "N/A" IF NOT APPLICABLE.

GRANDPARENTS OF THE BRIDE: _____

GRANDPARENTS OF THE GROOM: _____

PARENTS OF THE BRIDE: _____

PARENTS OF THE GROOM: _____

BRIDESMAIDS   USHERS

     ESCORTED BY: _____

     ESCORTED BY: _____

     ESCORTED BY: _____

     ESCORTED BY: _____

     ESCORTED BY: _____

     ESCORTED BY: _____

     ESCORTED BY: _____

     ESCORTED BY: _____

FLOWER GIRL   RING BEARER

     ESCORTED BY: _____

MAID/MATRON OF HONOR   BEST MAN

     ESCORTED BY: _____

BRIDE AND GROOM TO BE ANNOUNCED AS: _____

ABC MOBILE ENTERTAINMENT

WEDDING RECEPTION QUESTIONNAIRE

USING NUMBERS TO THE LEFT, PLEASE INDICATE THE EXACT
ORDER OF EVENTS. WRITE "N/A" IF NOT APPLICABLE
AND/OR "DJ" FOR DJ CHOICE.

____ TOAST   GIVEN BY: _____

____ BLESSING   GIVEN BY: _____

____ BRIDE AND GROOM'S FIRST DANCE: _____ (SONG)
                                        BEFORE/AFTER MEAL?

____ WEDDING PARTY'S DANCE: _____ (SONG)
                                        BEFORE/AFTER MEAL?

____ BRIDE & FATHER'S DANCE: _____ (SONG)
                                        BEFORE/AFTER MEAL?

____ GROOM & MOTHER'S DANCE: _____ (SONG)
                                        BEFORE/AFTER MEAL?

____ THROWING THE BOUQUET

____ GARTER REMOVAL AND TOSS

____ CAKE-CUTTING MUSIC? MODERN OR TRADITIONAL? _____

____ DOLLAR DANCE

ON A SCALE OF 1 TO 5 (5 = BEST), RATE EACH MUSIC CATEGORY

IN TERMS OF ITS LIKELY APPEAL TO THE <u>DANCERS</u> IN YOUR CROWD:

____ ROCK: ____ OLDIES: ____ MOTOWN: ____ BIG BAND: ____ DISCO:

____ 80s: ____ CURRENT DANCE: ____ ETHNIC: _____

WHAT (IF ANY) PARTICIPATION DANCES WOULD YOU LIKE?: ____

MACARENA: ____ ELECTRIC SLIDE: ____ CHICKEN DANCE: ____

OTHER: _____

____ DJ MEAL INCLUDED? PHOTOGRAPHER NAME: _____

VIDEOGRAPHER NAME: _____ IS THERE ANYTHING

SPECIAL THAT THE DJ SHOULD KNOW OR ANNOUNCE ABOUT THE

BRIDE AND GROOM, THEIR PARENTS, ANY OF THE GUESTS,

ETC.? _____

Sample Employment Agreement

ABC MOBILE ENTERTAINMENT

Street Address, City, State, Zip Code
Telephone Number, Cell Phone Number, Pager Number
E-mail address, Web site address

EMPLOYMENT AGREEMENT

This Agreement is made between ABC Mobile Entertainment, Inc., herein after referred to as COMPANY, and herein after referred to as EMPLOYEE. The COMPANY has established policies and standards to ensure the proper conduct of business and safety of its employees and clients. Employment with Company is subject to compliance with the following terms and conditions:

TERMS AND CONDITIONS

1. Employee shall fully comply with instructions of supervisors unless compliance threatens his/her safety or ethics.
2. Employee shall not falsify any Company paperwork.
3. Employee shall not deface or destroy Company's, client's, or fellow employee's property.
4. Employee shall not use or bring intoxicants or drugs to events or report to work while under the influence of alcohol or drugs.
5. Employee shall not leave equipment unattended at assignments, except to use the restroom or retrieve food or drink.
6. Employee shall not allow anyone other than authorized employees in work area(s) or in company vehicle(s).
7. Employee shall arrive at assignments 60 minutes prior to scheduled start time.
8. Employee shall not leave assignments until crew has arrived to dismantle equipment.
9. Employee requests for time off shall be made in writing and shall be permitted per managerial discretion in the form of a written memo. Requests must be made with at least 14 days notice except in the case of an emergency.

10. Employee shall adhere to Company dress code policy.
11. Employee shall periodically promote him/herself and mobile company at all mobile assignments.
12. Employee accepts responsibility for any and all materials, supplies, and equipment loaned and supplied during mobile assignments.
13. Employee shall relinquish any and all paperwork, materials, supplies, and equipment loaned belonging to Company, immediately upon dismissal from Company. Company shall retain any monies owed to Employee until said property is returned, replaced, or paid for.
14. Employee shall keep all Company paperwork, policies, procedures, pricing and marketing strategies and practices confidential and proprietary.
15. Employee shall keep work area(s) clean of glasses, bottles, ashtrays, and other clutter.
16. Employee shall not help him/herself to food or drink (except water) during an assignment, unless they are expressly offered by Client or facility representative.
17. Employee shall conduct him/herself at all times in a professional and courteous manner at bookings.
18. Employee shall follow any and all paperwork protocol and verbal or written assignments and/or instructions given by Company.
19. Employee shall receive work assignments approximately one week prior to events.
20. Employee shall attend all scheduled meetings and trainings.
21. Employee shall accept on a rotating basis, "on call" assignments as herein described: For all on-call assignments, between the hours of 6:30 P.M. and 10:30 P.M., Employee shall remain near a pre-agreed, constantly open telephone number, have functional car available, and be dressed in work attire. If Employee receives a call to work, he/she shall proceed immediately to complete the designated assignment. If originally scheduled Employee arrives at assignment, on-call Employee will immediately return to his/her on-call post after calling and alerting the on-call manager.
22. Employee shall be monetarily accountable for damage incurred to Company property should the damage occur as a result of Employee's negligence.

23. Employee shall receive a monetary penalty if through his/ her own negligence, the Company's contract with a client is in default.
24. Employee shall cease performance during an event in which he/she deems a real or implied threat of injury or harm to him/herself or Company equipment or music. Employee shall resume performance immediately after the client resolves the threatening situation.
25. Employee may deny any guest access to recordings or equipment during a performance.
26. Employee shall not perform independently or for an organization/company as a mobile disc jockey within a 60-mile radius of ABC Mobile Entertainment within a 5-year period of termination from Company.
27. Employee compensation and payment schedule for services rendered as a Mobile DJ Personality shall be provided in a written attachment separate from this Agreement.

EMPLOYEE understands that refusal or deliberate failure to comply with any of the above terms and conditions may be cause for disciplinary action ranging from a warning to immediate termination depending on the severity of the offense.

This Agreement cannot be canceled or modified except in writing by the COMPANY or EMPLOYEE.

The EMPLOYEE'S signature below indicates that he/she is requesting work as a Mobile Disc Jockey Personality for COMPANY, and has thoroughly read, understands, and unconditionally agrees to all of the terms and conditions listed above:

Employee Name: _____
Print

Employee Signature: _____ Date: _____

Social Security Number: _____

On behalf of ABC Mobile Entertainment:

_____ **Date:** _____
John Doe, Owner

Sample Bar/Bat Mitzvah Reception Questionnaire

BAR/BAT MITZVAH RECEPTION QUESTIONNAIRE

CLIENT NAME:_____ MITZVAH DATE:_____

PLEASE MAIL/FAX BACK THIS QUESTIONNAIRE TO US
NO LESS THAN ONE MONTH BEFORE YOUR AFFAIR!

CLIENT INSTRUCTIONS:

This *Questionnaire* contains several categories in which you'll be asked to fill in appropriate information. Here are some explanations to help make completing this sheet easier.

Initial Information: Include names of immediate family members, including step-parents if applicable and how you would like them announced.

Grand Entrance: Include your names (as host and hostess), names of siblings and the guest of honor. Choose a song to accompany each person walking in. You can choose different songs for each person or one for the entire group.

Candle Lighting: Include the names of people who will be coming to the cake. Write the names as your child calls them (ex. Nana Rose and Poppy Fred, Aunt Carol, Uncle Steve, Cousins Amy and Jason, etc.) The usual order for candle lighting is grandparents, aunts, uncles, cousins, older relatives, younger relatives, parent's friends, child's friends, parents, siblings, and Bar/Bat Mitzvah. The usual amount of candles is 14 (13 candles + 1 for good luck!) Try to group relatives and friends together to keep the amount of candles to 14. If you would like to include a memory candle to acknowledge a deceased loved one, you can also indicate that information on the sheet as well.

You'll also need to choose songs to coordinate with each candle. You may choose all Jewish music, any other song (preferably upbeat), or a combination of both. You may want to match a specifics song to each person or group of people lighting the candles. If you don't have specific preferences for songs, our DJ/MCS will choose the upbeat music for you to complement each candle.

Hora: Indicate which immediate family members you would like to be lifted in the chair during the Hora.

Motzi, Toast, Kiddish: Indicate which guests will be performing theses honors. Please note that a Kiddish is sometimes not performed at the reception. Please ask your caterer/banquet manager.

Special Announcements: If appropriate, indicate if there are any special announcements about any guests that you would like announced (ex. birthday's anniversaries, engagements, etc.)

Special Dances: Please indicate which songs you would like to hear for the host/hostess Dance, Bar/Bat and Friend Dance (if appropriate) and Bar/Bat and Parent Dance. Your DJ/MCS will be happy to give you suggestions for these dances.

CONTINUED ON OTHER SIDE

PAGE 1

Figure 11.2 Bar/Bat Mitzvah reception questionnaire.

Sample Event Questionnaire

---

1) ARE THEIR ANY SPECIAL ANNOUNCEMENTS OR DEDICATIONS YOU WOULD LIKE YOUR DJ/MCS TO MAKE?
___ YES ___ NO. IF YES,

a) (SONG TITLE & ARTIST)_____
(DEDICATED TO)_____
(MESSAGE)_____
b) (SONG TITLE & ARTIST)_____
(DEDICATED TO)_____
(MESSAGE)_____
c) (SONG TITLE & ARTIST)_____
(DEDICATED TO)_____
(MESSAGE)_____

2) HOW MUCH ENERGY AND CROWD INTERACTION WOULD YOU LIKE FROM YOUR DJ/MCS?
(RATE 1-10, 10=MOST):_____

3) TYPES OF MUSIC YOU WOULD MOST LIKE PLAYED (RATE 1-10, 10=MOST):
_____ OLDIES/MOTOWN _____ CLUB/HIP-HOP/R&B _____ CLASSIC ROCK _____ 70's & 80's DANCE
_____ BIG BAND/SWING _____ ALTERNATIVE _____ OTHER _____

4) PHOTOGRAPHER NAME/COMPANY:_____

5) VIDEOGRAPHER NAME/COMPANY:_____

6) WILL MEALS BE PROVIDED FOR YOUR TWO DJ/MCS?_____

7) LIST "MUST PLAY" SONGS FOR THE DANCE PORTION OF YOUR RECEPTION (SONG TITLE & ARTIST):

             SONG TITLE   ARTIST

_____
_____
_____
_____
_____
_____
_____
_____
_____
_____
_____

CONTINUED NEXT PAGE

Figure 11.2 *(Continued)*

Sample Event Questionnaire – cont'd

Initial Information
Bar/Bat Mitzvah_____
Bro/Sis_____
Parents_____
Grandparents:
Mother's_____ (what child calls them)_____
Father's_____ (what child calls them)_____

Grand Entrance   Name   Song
Host/Hostess_____
Bro/Sis_____
Bro/Sis_____
Bro/Sis_____
Bro/Sis_____
Guest of Honor_____

Candle Lighting
            Name   Song
_____
_____
_____
_____
_____
_____
_____
_____
_____
_____
_____
_____
_____
_____
_____

Hora: To be lifted in chair_____

Toast_____ Motzi_____ Kiddish (if app.)_____
Special Announcements_____

Special Dances
Host/Hostess_____
Bar/Bat & Friend_____
Bar/Bat & Parent_____

Figure 11.2 (*Continued*)

Sample Bar/Bat Mitzvah Games and Dances Information Sheet

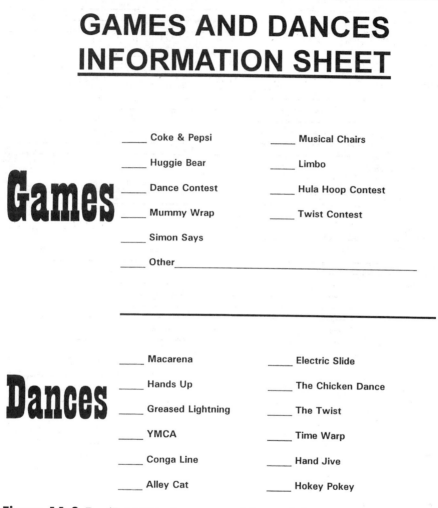

GAMES AND DANCES
INFORMATION SHEET

Games

_____ Coke & Pepsi   _____ Musical Chairs

_____ Huggie Bear   _____ Limbo

_____ Dance Contest   _____ Hula Hoop Contest

_____ Mummy Wrap   _____ Twist Contest

_____ Simon Says

_____ Other_____

Dances

_____ Macarena   _____ Electric Slide

_____ Hands Up   _____ The Chicken Dance

_____ Greased Lightning   _____ The Twist

_____ YMCA   _____ Time Warp

_____ Conga Line   _____ Hand Jive

_____ Alley Cat   _____ Hokey Pokey

Figure 11.3 Bar/Bat Mitzvah games and dances information sheet.

Sample Bar/Bat Mitzvah Candle Lighting Suggestions

MEMORY
- ☐ Dodi Li
- ☐ I Will Always Love You
- ☐ Memory
- ☐ Schindler's List
- ☐ Somewhere Out There
- ☐ Tears In Heaven
- ☐ Unforgettable

FRIENDS
- ☐ ABC
- ☐ Celebration
- ☐ Everybody Have Fun
- ☐ Friends (Bette Midler)
- ☐ Girls Just Want to Have Fun
- ☐ I'll Be There For You
- ☐ Lean On Me
- ☐ A Little Help From My Friends
- ☐ Still A Friend Of Mine
- ☐ Thank You For Being A Friend
- ☐ That's What Friends Are For
- ☐ Together Forever
- ☐ We Are The World
- ☐ We Go Together
- ☐ Whenever I Call You Friend
- ☐ You're My Best Friend

GRANDPARENTS, AUNTS & UNCLES
- ☐ As Time Goes By
- ☐ In The Mood
- ☐ Simon Tov
- ☐ Sunrise Sunset
- ☐ Times Of Your Life
- ☐ To Life
- ☐ Tradition
- ☐ Unforgettable
- ☐ Wonderful World
- ☐ Young At Heart
- ☐ I Just Called To Say I Love You
- ☐ I Only Want To Be With You
- ☐ I Say A Little Prayer
- ☐ Let The Good Times Roll
- ☐ Let's Hang On
- ☐ Respect
- ☐ We Are Family
- ☐ Where's The Party?
- ☐ Stand By Me

SISTER & BAT MITZVAH
- ☐ Girls Just Want To Have Fun
- ☐ I Saw Her Standing There
- ☐ I'm Every Woman
- ☐ I'm Too Sexy
- ☐ Isn't She Lovely
- ☐ Material Girl
- ☐ Pretty Woman
- ☐ Uptown Girl
- ☐ We Are Family
- ☐ Brown Eyed Girl

Figure 11.4 Bar/Bat Mitzvah candle lighting suggestions.

Sample Bar/Bat Mitzvah Candle Lighting Suggestions – cont'd

BROTHER & BAR MITZVAH
- Get Ready For This
- Havin' A Party
- He's So Fine
- I Wanna Be Rich
- I'm Too Sexy
- Macho Man
- Rocky Theme
- Sharp Dressed Man
- Soul Man
- We Are The Champions

NAMES
- Angie
- Barbara Ann
- Bette Davis Eyes
- Billie Jean
- Copacabana (Lola)
- Danny Boy
- Daniel
- Gloria
- Help Me Rhonda
- Hit The Road Jack
- Johnny Be Good
- Jumpin' Jack Flash
- Leroy Brown
- Louie, Louie
- Maggie May
- My Sherona
- Peggy Sue
- Rockin' Robin
- Roxanne
- Run Around Sue

MOM & DAD
- Can You Feel The Love Tonight
- Endless Love
- Everything I Do
- Have I Told You Lately
- I Got You Babe
- I Just Called To Say
- I Want To Hold Your Hand
- I Will Always Love You
- I'll Be There For You
- Sunshine Of My Life
- Time Of My Life
- We Are Family
- We've Only Just Begun
- When I Fall In Love
- Wind Beneath My Wings

OUT OF TOWN GUESTS
- Boy From New York City
- California Dreamin'
- California Girls
- Carolina On My Mind
- Caribbean Queen
- Freeway Of Love
- Georgia On My Mind
- Go West
- I Left My Heart In San Francisco
- Rhinestone Cowboy
- Life Is A Highway
- My Kind Of Town
- New York, New York
- Sweet Home Alabama
- Yellow Rose Of Texas

Figure 11.4 (*Continued*)

Sample Bar/Bat Mitzvah Candle Lighting Suggestions – cont'd

<u>*JEWISH*</u>

- ❏ Dayena
- ❏ Dodi Li
- ❏ Erev Shel Shoshanim
- ❏ Hava Nagilah
- ❏ Hora Medley
- ❏ Jerusalem Of Gold
- ❏ Kinena Hora
- ❏ Oseh Shalom
- ❏ Shalom Alechem
- ❏ Simon Tov
- ❏ Sunrise Sunset
- ❏ To Life
- ❏ Tradition
- ❏ Tzena-Tzena
- ❏ Yaseh Shalom

<u>*SMALL CHILDREN*</u>

- ❏ Go Go Power Rangers
- ❏ Hakuna Matada
- ❏ I Love You (Barney Theme)
- ❏ Just Can't Wait To Be King
- ❏ Mr. Roger's Neighborhood Theme
- ❏ Rainbow Connection
- ❏ Sesame Street Theme
- ❏ Zip-A-Dee-Doo-Dah

Figure 11.4 (*Continued*)

The author of *The Mobile DJ Handbook* is not legal counsel or an attorney and is not offering legal advice by providing sample contracts or forms for you to copy. It is suggested that you seek professional legal counsel prior to using any contract or form to ensure its validity in the state in which your business operates.

*A true leader is not the one
with the most followers,
but one who creates the
most leaders.*

NEALE DONALD WALSCH, Author
Conversations with God

12

Records of Success

Define success on your own terms. Don't let others make the rules for your happiness. Often it's a balance of family life, personal achievement, and doing good for others. Don't let others' "shoulds" and "musts" rule your choices. Make up your own mind and set the course for your own life.

SID VANDERPOOL, DJ ENTERTAINER
Editor, *DJzone.com*, on-line DJ magazine
Twin Falls, Idaho

Interviews with America's Top DJs

The following are interviews I conducted with some of America's most successful and respected DJ entertainers and business owners.

How did you get started as a mobile DJ?

J. R. SILVA
DJ, VJ, and KJ. Performer and producer with *SILVA Entertainment* in Orlando, Florida

    I was working as a part-time DJ at a small-market radio station. I had a friend who owned a mobile disc jockey service and saw him at a party. He had a great sound system and music, and people booked him for those things alone. But he said that he needed help because he didn't really like to talk on the mic. On many occasions I would go out with him to do the voice work, and he would play the music and run the board. Business got so busy that I decided to buy my own equipment so that he could subcontract me out for

Figure 12.1 DJ J. R. Silva. Courtesy *DJ Times* magazine.

bookings. We worked this way between 1987 and 1989. Finally, I decided to go out on my own, because he had some DJs working for him that weren't very professional, and I wanted more customer interaction. I got a small office and started doing two shows every weekend. This quickly grew to five. Since going out on my own, I have always striven to make the industry more legitimate.

RICHARD HART

Founder, Hart to Hart Entertainment, Long Island, New York

About 15 years ago, because my father was a bandleader, I used to dance with the crowd at parties. I would grab the microphone and motivate the people with different dance steps. Then I started to use DJ equipment along with the band. Eventually I was doing parties myself. I worked the entire room with a wireless mic while singing and dancing along with the crowd. The roadies who handled the equipment started dancing along with me and then became emcees. The response was so enormous that suddenly a hobby was turning into a business. Things really took off in 1987. By 1990 we were doing 40 jobs in a weekend and had 20 crews.

Figure 12.2 DJ Richard Hart. Courtesy *DJ Times* magazine.

"KEVIN HOWARD" ST. JOHN

President, *The Howard Group, Inc., d.b.a Sounds Unlimited*, Seattle, Washington

Beginning in 1975, I worked my way through high school and college as a mobile DJ. While in college, I saw the opportunity to bring some real business value into what was then, in my area, a fledgling industry without a lot of marketing savvy.

"RANDI RAE" TREIBITZ ("THE MITZVAH MAVEN")

President, *Randi Rae Entertainment*, Fords, New Jersey

I used to be a graphic artist in the advertising business. I also created promotional ideas to sell advertising to nightclubs, restaurants, and hotels. I would coordinate the entire show or contest, and then would act as the emcee for the event. Through one of my accounts in the mid-80s, I met a disc jockey with great technical skills, and we started working together part-time. He worked behind the console, and I was in front of the crowd. Eventually my

Figure 12.3 DJ Kevin Howard.

partner wanted to get out of the business. We were able to work out a deal where I got the equipment and music, and I continued to work on my own part-time for a while. Then I worked for New Jersey's most successful DJ company for a few years. After that, I made the plunge into my own mobile DJ business full-time.

BOB DEYOE

President, *Desert DJs*, Tucson, Arizona. Featured moderator at *DJ Times'* International DJ Expo. Recipient of *DJ Times'* 1999 International DJ Expo "DJ of the Year for the Best New Game" category award. Recipient of 1997 *DJ Times'* International DJ Expo "Top Ten DJ Entertainers" award. Recipient of *DJ Times'* International DJ Expo "Top Ten Wedding DJs" award in 1996.

Figure 12.4 DJ Randi Rae.

From the time I was 6 or 7 years old, I always wanted to be a disc jockey. In 1979, when I was a senior in high school, I got my first professional job for a college graduation. I used the equipment I had at home. I think I charged them $60.00, and it cost me $120.00 to do the job! From then on I worked as a mobile DJ on the side until eventually I went full-time.

JEFFREY GREENE

President, *Party Time DJs*, Davie, Florida. Winner of national awards and recognition in the international DJ community including "Best Party DJ in America."

Figure 12.5 DJs Bob Deyoe and Jennifer Payne from Desert DJs. Photo Courtesy of Daniel Snyder Photography.

I loved to dance and always found myself being at the center of the dance floor leading dances and such. A DJ at a party I attended liked how I motivated the crowd and asked me if I'd like to learn to be a DJ. After working with him at a couple of parties I knew I'd found my "calling," so to speak. Now I own and operate a large multi-system operation. We operate four regional offices, and we also offer photography and video services.

ROXANNA GREENE ("THE *MACARENA* QUEEN")

President, *A Woman's Touch Productions/Roxanna Greene Entertainment & DJs*, Hialeah, Florida. 1996-1997 Recipient of *DJ Times'* International DJ Expo "Entertainer of the Year" award. Mobile Entertainers Guild of America award recipient for teaching the *Macarena* to the DJs of America. Founding board member of USMEA.

My brother Jeff and I used to enter dance contests in the '70s and we were on "Dance Fever." In 1979, Jeff started Party Time DJs. One day, he got a wedding booking in Key Largo, and he asked me if I would help out with the bride and equipment. We got a

Figure 12.6 DJ Jeff Greene. Courtesy *DJ Times* magazine.

rave review from all the guests for the interactive dancing we were doing, so we knew we had a good thing going.

How can someone excel as a professional mobile DJ?

RANDI RAE

Act professionally! Whether you are making $100 or $1000 on a job, always give the same high-quality performance. Take your job seriously.

Figure 12.7 DJ Roxanna Greene, president, A Woman's Touch Productions/Roxanna Greene Entertainment & DJs, Hiahleah, Florida.

KEVIN HOWARD

You need to find a niche—an area of service that is not being satisfied by the DJ companies in your area. That may mean offering high-end full productions, or choosing to tap into the low-end economy market. It's a matter of positioning. Over the past 20 years the level of skill of mobile disc jockeys has improved dramatically. The trend is definitely moving toward a requirement for DJs to become entertainment hosts who can motivate an audience to dance and participate in activities.

JEFF AND ROXANNA GREENE

Recognize that you are in a luxury industry that has finally begun to mature. If you want to go beyond being a human jukebox, then provide an asset to people's life events by being a professional entertainer. Constantly strive to be a better businessperson, and always act ethically. This includes buying music through legal sources. Incorporate your business and carry liability insurance.

BOB DEYOE

Never stop learning. Leave your ego at home. Network with other professionals in the business. Devour as much information as possible, like this book!

J. R. SILVA

You excel as a mobile DJ by providing excellent service and being professional. You need to know the music and how to have fun. People want to party with someone who knows how to have fun. It helps if you embrace the standard party songs and party-host techniques. I'm always looking for a new twist to an old song, or for something different to do with the music. I am like a sponge, willing to soak up and learn new dances that I can teach at events or with my staff. It's much better to come over-prepared to a party than not having enough in your "bag of tricks."

RICHARD HART

They have to be sincere and really enjoy what they're doing. You can't fake enthusiasm.

How can a DJ select the right music for the crowd?

KEVIN HOWARD

You determine the probability of what your audience will enjoy based on socioeconomic and demographic factors. You combine this determination with the requests from your client and audience. Then you begin a period of trial and error by providing the music you believe the crowd will like and seeing how they respond.

JEFF AND ROXANNA GREENE

You meet with the client and ask for the average age of the guests attending the party and for information pertaining to ethnic background. When doing a corporate affair, we ask if the event is being held for upper management or for the workers. If it's an event for the workers, we ask for an ethnic breakdown of the guests. These questions are asked so we will know what kind of music to bring and play. We not only ask a client to tell us what music they do want, but also what music they don't want. We also encourage guests to make requests at the party. Lastly, we add the songs to the client and guest requests that, through experience, we know are the right songs to play.

J. R. SILVA

Consulting with the client beforehand is really important. Find out the average age of the guests. If you're able to mingle with the crowd during the social hour, you can ask them what their music needs are. You also need to use your intuition by feeling the vibes in the room and listening to the nuances. I also ask the customer directly for the tone they are trying to set. Some clients want a high-energy presentation. Others may want the DJ to be more relaxed. This information is important to note in the preliminary consultation. And I also ask, "How much enthusiasm do you want from your party host, on a scale of 1 to 10?"

RANDI RAE

By talking with the client before the event, I can find out what they want and don't want played at their affair. At the party, I take requests and always play the radio versions of songs. I like to play four to six songs in a set and break up the sets with two slow songs. The first slow song I play is from the era or type of music from the set I just completed. The second is from the era or type of music in the set I am about to play. I have music from the 1920s to today, and a lot of ethnic music as well.

BOB DEYOE

A DJ can select the right music for their crowd by collecting various forms of feedback. We send a contract to our clients in which they can select types of music and choose specific songs they want from a song list. During our disc jockey staff meetings, everyone shares the top 10 songs in every music category that works best for them. At the event, a DJ also has to be able to "read" the crowd, by taking demographic factors into account.

RICHARD HART

There are certain party songs that will never go out of style and work to motivate every crowd. You could play *Shout* at every party and it would work. You also have to honor the requests of the people who hired you. The rest is "reading" the crowd. This involves having extensive knowledge of the music listened to by the age group for

whom you are playing. Pay attention to what the group is responding, and if something isn't working, change it. You don't want to
lose the crowd on the dance floor. We do dance routines that people
can follow and strongly believe in the interactive approach. This
makes the songs come alive and keeps people on the dance floor.

What are some good crowd "icebreakers?"

RANDI RAE

Group participation dances. I try to balance my dance floor with
a combination of participation dances and regular dancing. It all
depends on what I see in my crowd and what the client wants.
I have a dance for everything and could fill up 4 hours with nonstop
participation dances for every type of music. Any DJ can learn these
dances by renting videos and practicing at home. If the party you are
doing has a special theme, work the theme into the entertainment.

KEVIN HOWARD

With weddings, you are blessed with traditional activities that, if
used properly, will allow you to bring the entire audience together.
With other events, you need to be more proactive. Some examples of
icebreakers we use are snowball mixers, where you keep increasing
the number of people on the dance floor by having those dancing
bring out others every 20 seconds or so. At a corporate event, I ask
the planner to give me the names of 6 to 12 of the most outgoing
people attending, and then I "volunteer" those people to be the first
ones up dancing. Of course, contests and participation dances are
also good icebreakers.

J. R. SILVA

We tend to use *The Twist* or participation dances in the first half-
hour. If I feel the audience is receptive, or the client has mentioned
they want follow-me songs like: *The Chicken Dance* or *The Hokey
Pokey*, then we tastefully lead those dances. In most situations we
have already identified with the client which songs are most suited
for their event. We have over 50 different dances or icebreakers we
can use, so we can stay away from anything the client or wedding
couple has deemed overplayed or corny. We try to incorporate the
guests' requests as much as possible. Our goal is to stay one step
ahead of the audience, leading and building the party energy up to
meet the client's expectations.

RICHARD HART

You can ask a lot of questions over the microphone and get people to make noise. Audiences respond to your saying things like, "Is everybody ready to have a good time?" and "Let me hear you scream." If I don't like the response I'll even repeat myself. Get silly. Get crazy. It will put a smile on people's faces. Put hats and sunglasses on guests when they're out on the dance floor. I also use the guest of honor to get people going. For example, at a Bat Mitzvah I might say, "Ladies and gentlemen, how many people are ready to party with Jennifer? Does Jennifer look outrageous out here? Let me hear you scream for her!"

BOB DEYOE

There are a number of different things I use. An icebreaker for weddings is the "Marriage Countdown." I use participation dances, such as a conga line with a "follow me" sign on the back of the leader.

JEFF AND ROXANNA GREENE

Be a crowd motivator. You can break the ice with the "King of Conga" routine, *Macarena*, the Electric Slide, Dollar Wine, YMCA, the Italian Chicken, the German Chicken, or many other participation dances. If you have a solemn group, you've got to encourage them to make requests. You can do this by walking around and mingling with the guests during cocktail hour and the meal. If you're doing a corporate event where there are mostly men, don't do participation dances that require a female partner.

What inspired you to start your own DJ service?

BOB DEYOE

Many years ago, I got a call for two events on the same night. I only had one system so I knew it was time to expand. I've always been inspired to be a mobile DJ. It is the only thing I've ever wanted to do.

JEFF GREENE

I was at a party where there was a DJ. He saw me dancing and my interest in the music, and invited me to come and do a couple of events with him. It worked out really well.

KEVIN HOWARD

I had been a mobile DJ for a long time, and I just saw a true need for excellent service. I had a crystal-clear goal and knew exactly what I wanted to provide. This included customized music, entertainment using top-quality equipment, and energetic, professionally trained DJs, and a product that was consistent and wasn't available in the marketplace. I started out with that goal in mind.

J. R. SILVA

Entertainment is constantly changing. I've always tried to be the leader in the market by being the first to bring in new services. I attend the DJ Expos and network with other disc jockeys for information. I pay attention to the customers' needs and always explain how we can not only meet their needs but surpass them. I also have a terrific staff who understands the responsibilities that come with the work we do. When I train someone, I have him or her follow my lead and do things my way for a while until he or she is up to speed. Then, he or she can start adding their own creativity to the mix. This way, we can all learn things from one another.

How have you set up your business operation?

KEVIN HOWARD

We're a corporation. I have a general manager, several sales people, a warehouse manager, and a whole slew of disc jockey emcees and set-up people. Our corporation is an entertainment production company with different divisions. We have a fantasy casino division, a decor and decoration division, and three distinct disc jockey companies. We are established in Washington and Oregon with 16 full-time employees. In any given quarter, we have 60 to 100 people on payroll.

J. R. SILVA

In the California location, I have one person who serves as my office and sales manager, a second who is my DJ manager, and a third who is my warehouse manager. That trio is able to keep me in touch with how the business is going. The DJ and warehouse manager both work part-time; the sales/office manager works full-time. I have 10 employees who work exclusively for me. In the Orlando market, I manage the DJs, the office, and the sales. I have a roster of

6 to 10 full-time DJ employees, and I hire additional DJs, dancers, and lighting subcontractors during peak season. I subcontract out the warehouse work.

RICHARD HART

I have high-energy people working for me. I try to get the most from those people by hyping them up as much as possible and by keeping their enthusiasm going week after week after week. In addition to running great parties, I always have to remember that I'm also running a business. I have computers, great sales people, and I treat my customers well.

BOB DEYOE

We have a business front office of about 1200 square feet in Tucson, Arizona. We also have two additional offices and a warehouse that share the same space. In addition to my wife and me, we have 12 DJs who are employees, an assistant manager, and an assistant office manager. We use a subcontractor DJ when we get overbooked.

JEFF GREENE

I have four offices: a main office, one at a banquet facility, and two sales offices. Each is fully computerized and run by an office manager. We also have a sales staff and secretary at each location. Right now we have a total of 30 employees.

ROXANNA GREENE

I have one sales location and also work from my home office.

RANDI RAE

I have one location and several systems that my subcontractors use. I handle every aspect of my business myself. I always go out on a job as part of a two-person team. The DJ plays the music, which I program, and I act as the emcee. I perform mostly in New Jersey, but I have also entertained in California, Connecticut, Georgia, South Carolina, and Florida. By attending the DJ expos, I have made a number of contacts around the country. When I get a booking outside my usual business area, I call a contact near the booking location and use that company's system and disc jockey.

Why has your business been successful?

RICHARD HART

Because I know how to run a job correctly and am able to really motivate a crowd and get them involved. I don't just play music. I deal well with the clients and the caterers. I love what I do, and I have a good time doing it.

J. R. SILVA

I've always tried to be the leader in the market by being the first to bring in new services. I attend the DJ Expos and network with other disc jockeys for information. I pay attention to the customers' needs and always explain how we can not only meet their needs but surpass them. I also have loyal, honest employees who are excellent entertainers.

KEVIN HOWARD

We were the first in our area to establish a professional approach to the industry, and we expanded rapidly. In 3 years we became the largest company in the Seattle area. We also have excellent employees and commit ourselves to delivering what we promise. We operate with integrity, courage, and honesty. Referrals have always come as a result of these practices.

RANDI RAE

I believe my success is due to personality, caring, and a total commitment to my clients. I work 20 hours a day and would never change my career—unless someone offered me a talk show. I love performing in front of people.

JEFF AND ROXANNA GREENE

Although we operate separate businesses, each of us has always striven to be the best. We believe very heavily in running our businesses with a "hands-on" approach and are very aware of what the competition is doing. I [Jeff] also believe in constant training to improve performance. That's why I conduct in-studio trainings for my staff.

BOB DEYOE

I think part of it was the timing. We made a huge commitment about 10 years ago to be number one. There was a company about three times our size that was very well known. Our goal was to emulate them and do what they did even better. It took a couple of years, and they "blinked." Then we caught up to them, and they sold out. The new owners had no chance of catching up to us. We have excellent equipment and talented people whom we treat really well. I have strong marketing ability, and my wife is able to create marketing materials from my ideas.

How do you sell and market your services?

BOB DEYOE

A small Yellow Pages ad, a three-color brochure that includes inserts and copies of articles and interviews that have been done with us. We like everything we send out to look really, really sharp. Clients are invited to stop by the office and visit us. We have six videos to view; two were specifically made for bridal clients and walk them through the bridal party introductions.

KEVIN HOWARD

How don't we! We use everything that's available to us, including newspaper and magazine advertising, bridal guides, dominant Yellow Pages advertising, radio co-op, trade shows, networking groups, trade associations, direct mail, and telemarketing. We also produce five bridal shows yearly, including the largest one in our state.

RANDI RAE

My entire business is referrals. I do bridal shows where I am the only DJ and I'm able to perform live. I have a "2-foot rule." Anyone I come within 2 feet of, during the day, is a potential client and therefore should know what I do for a living. I do a lot of public relations work during a party by shaking hands and meeting people.

JEFF AND ROXANNA GREENE

Word of mouth would have to be number one. We only use the free business listing in the Yellow Pages. We used to be the biggest advertiser, but now we do a hell of a lot more business than we did then. That's got to tell you something. Today we advertise in direct marketing sources. For example, wedding and Bar Mitzvah shows, special events, and advertising guides. Performing on stage at showcases allows many prospects to preview our services at one time.

J. R. SILVA

A lot of it is good old-fashioned networking. I've always relied on personal consultations, great brochures, and targeted newsletters to different markets. I've also used video a lot, either in-person or by mailing it to a potential customer. Video is a fantastic tool for showing a customer exactly what they're buying, and I have specialized videos for weddings, proms, Bar and Bat Mitzvahs, etc.

RICHARD HART

We use videotapes to show people what we do. We go into different types of showcases. We establish a good rapport with caterers and people who plan parties. We advertise and have a sales staff who sells our product.

How much of your annual budget is dedicated to marketing and advertising?

JEFF AND ROXANNA GREENE

Too much, roughly 15 to 20 percent.

KEVIN HOWARD

10 to 20 percent.

J. R. SILVA

I spend about $5,000 a year. Even though I want to do more, and have marketing campaigns that I plan to further develop, I try to keep my advertising budget simple. Once you have a balance with how much is required to reach new customers and still take care of the important recurring customers, then you can focus on new markets and goals. You can get caught up in all this marketing, then you're stuck needing to do more volume just to pay for the advertising . . . that's just crazy. The Internet has changed the marketing field as well. Yellow Page advertising is dead as brides jump on-line to find their entertainment. Our website is a constant work in progress because I'm always trying to make it better each month. In the future, I intend to make Funtalent.com my virtual office on-line.

RICHARD HART

About 5 percent.

BOB DEYOE

About 6 percent of gross. If I were brand-new, it would have to be a lot more—20 to 30 percent.

How should a new DJ company price their services?

BOB DEYOE

We did research in our area when we first started and priced ourselves 10 to 20 percent below the competition. That was the draw, initially, for people who had never used our service to try us out. We knew they would be ecstatic. Through the years, we slowly changed our rates to be the same as the competition, and then eventually made them a little bit higher. If you can find a way to see what the competition is doing out there, do so. Ask the catering managers and party planners for their feedback. Know what your talent is and find a niche in the marketplace.

RANDI RAE

Price your services competitively, based on your experience. Do not overprice yourself. However, don't low-ball your price just to get a job. That practice does the DJ industry an injustice, not a service.

RICHARD HART

First you have to determine what type of parties you want to do and what markets you want to be involved with. Once you've decided on these things, you have to find out what the competition is charging and adjust your prices accordingly.

KEVIN HOWARD

You need to find a position in the market and occupy it. Make a choice as to whether you are going to offer a low-end service or a premium service. People buy based on price, quality, and service. They can have two of those three.

J. R. SILVA

Back in the nineties I think the average rate for most DJs was around $100/hr. Fortunately the rates have pushed up since then. In Florida, wedding DJs are currently getting between $595 and $1295 for 4-hour events. DJs are offering either more add-ons or more showmanship to get their price up and to add value to their shows. Find a niche in the marketplace and price your services accordingly. I charge a couple of hundred dollars more than the competition because I know we're just that much better. I want to use all of our experience and creativity and get our asking price. Nationally, the average rate is about $100 an hour for a professional disc jockey.

JEFF AND ROXANNA GREENE

You need to decide whether to establish your business as a full-time or part-time operation. You have got to put together a budget and know your costs. Then, charge the highest dollar amount that you believe you can get. For example, if every DJ service charges $400, and you know that you're good and you can entertain beyond that, perhaps you should charge $600. At first, you may not get as many jobs, but at the jobs you do get, people will really appreciate what you do and will refer you more often. There is a certain segment of the population that believes that the highest price means the best service. You know what? They're right.

How do you choose your equipment and where to buy it from?

J. R. SILVA

When I was starting out in a small market, there were only music stores. I relied on those stores to tell me what equipment I needed. A couple of years later, after I had a few systems, I discovered there were DJ stores that advertised in the trades and could sell me more industrial gear. Magazines like the *DJ Times* give you a good idea of what is available on the market. You have to set aside a budget and know what will best serve your needs. You'll need to consider whether you will want to have top-of-the-line equipment or settle for something more middle of the road. If you are looking for outrageous sound, you might buy big speakers and tri-amped sound systems. If you want something lighter, you go might go for the self-powered gear, which is much easier to haul around.

Some business owners don't buy equipment at all and just rent gear when they need it.

JEFF AND ROXANNA GREENE

Research the product that's out there. Try it out for yourself if you can. Ask your peers about its reliability. Buy what you like, and take into consideration how you will transport the equipment. In larger markets, you can shop retail and play one store against another for price and service. In smaller markets, you may want to consider mail order.

KEVIN HOWARD

We buy only state-of-the art equipment. It's important to consider if your purchase will increase the entertainment value for your client. The tools that truly enhance an entertainer's ability to motivate an audience are a necessity for the true professional. I think many people spend too much money on "DJ toys." Buy the stuff that makes the difference for a good performance, like a quality CD player and a good quiet mixer. I recommend buying mail order in volume.

BOB DEYOE

Check the buyer's guide in trade publications. Although some people don't have a lot of money to spend when they are first starting out, it is important to put money into your equipment. Otherwise, you may have to repair your gear down the road. Buy equipment that is professional, durable, and transportable. Whomever you buy from should be able to repair it. Mail order is fine, but if something goes wrong, hopefully you will be able to fix it rather than throwing it away.

What are the keys to success and prosperity with a mobile DJ service?

RICHARD HART

You've got to work your business 12 hours a day. You have to be able to motivate a crowd and get a high from making the crowd enjoy themselves. You also have to give your business 110 percent. You have to live it, breathe it, and treat it like your very own baby. Every party has to be great. If you do 99 great parties and 1 bad party, it seems like the whole world knows about the bad one. No matter how big or how small a job is, every job has to be perfect. Every customer has to be handled with proper care from beginning to end. The customers are educated these days, and they know that it's the entertainment that makes the party.

BOB DEYOE

It's best to be very ethical and to treat all of your clients fairly. If somebody is not happy with a job you did for them and you don't respond, it will come back to haunt you. The only companies that have ever done well in our area are those that market themselves and continually shake hands with every new catering director and party planner. You have to be able to back up what you say with really good DJs. If you want to be successful, you need a really good product. Your disc jockeys have to be well trained. Let everyone in town know you're absolutely the best and you'll guarantee it! If you want to get into this business, go into it for the right reasons. Do it because you want to do something really well. Our mission statement talks about providing clients with the best possible service, and that's why we're here. We would certainly like to make a profit, but we also want to provide the best possible service.

RANDI RAE

Run a legitimate business and don't be greedy. Learn the business and be a professional. Start small and do not overbuy. Stay focused. Treat your DJs well. Remember the client is always right. Bear in mind that when you are out doing a job, you're not just representing yourself, you're representing an entire industry.

KEVIN HOWARD

Put in a 60- to 80-hour week and work hard at becoming successful. You must position yourself correctly in the marketplace.

J. R. SILVA

You can succeed and prosper by being an incredible entertainer. Even when you're not in the mood to lead the YMCA, Electric Slide, or conga, do them anyway with great enthusiasm if you know it's going to please the audience. You must know when to inject a fun dynamic and when to just play music. It's also very important to be a team player when you are at the event. You'll be able to make good contacts and build your business leads.

Do you still personally go out on bookings?

RANDI RAE

Absolutely! I primarily do parties that use my expertise as a professional emcee entertainer. I also specialize in Bar and Bat Mitzvahs. The fee for my services depends on additions, such as props, prizes, dancers, lighting, video, and Karaoke, and runs from $650 to $3650. To generate additional revenue, I will also do party planning for a client.

KEVIN HOWARD

One of the subdivisions of our corporation is a DJ company that books only myself. I go out with a crew to do full production work. My average range is between $950 and $1300. The high end would be $3000.

J. R. SILVA

I average three to four shows a week DJ-ing for theme events, Karaoke, and video dance parties. You name it. Depending on who and what we bring, shows range from $450 to $2500.

RICHARD HART

I still go out and do parties every week. I normally charge $6000 for a full production with dancers, to give people the greatest day of their life. The client's happiness is my gratification.

BOB DEYOE

I perform all the time. I consider myself to be first and foremost an entertainer and secondly a manager of the company. I would never want to own a DJ company and not be one of the performers. I don't have a higher fee for myself than my other people. We offer a request fee to our staff. If a client requests a specific disc jockey, the request fee goes directly to them.

ROXANNA GREENE

As a party host entertainer, I get $125 an hour or $500 for a 4-hour event. I get $250 to $300 an hour for my custom package.

JEFFREY GREENE

You bet. My current rate is $375 an hour without lights or any extras. Many of my staff command $250 to $300 an hour and our production packages range from $2000 to over $10,000. If every professional, experienced mobile DJ in America set their prices higher, then the entire industry standard would be raised, and clients would simply have to accept the increase accordingly. Disc jockeys spend between 12 and 20 hours on an event. We don't take breaks at a performance, offer a larger variety of music than bands, and are often better entertainers. In addition, we spend thousands of dollars on our music collections, equipment, and training. My point? Those who have invested in being the best deserve to be paid the highest rates because they've earned the right!

Appendix

Accessories/Cables/Connectors/Parts/Stands

- American DJ—www.americandj.com
- Atlas-Soundolier—www.atlas-soundolier.com
- Canare—www.canare.com
- Cord-Lox—www.cord-lox.com
- Decibel—www.decibelproducts.com
- Eupen—www.eupen.com
- Furman—www.furmansound.com
- Gemini Sound Products—www.geminidj.com
- Gepco—www.gepco.com
- Grundorf—www.grundorf.com
- H&F Technologies—www.audio2000s.com
- Hosa Technology, Inc.—www.hosatech.com
- Lightning Masters—www.lightningmaster.com
- MBT Lighting & Sound—www.mbtinternational.com
- Middle Atlantic Products, Inc.—www.middleatlantic.com
- Neutrik—www.neutrikusa.com
- Parts Express—www.parts-express.com
- Perfectdata—www.perfectdata.com
- Planet Waves—www.planet-waves.com
- Pro Co—www.procosound.com
- Raxxess—www.raxxess.com
- Sony—www.sel.sony.com
- Switchcraft—www.switchcraft.com
- Ultimate Support—www.ultimatesupport.com
- Whirlwind—www.WhirlwindUSA.com
- Zercom—www.nortechsys.com

Books/Publications

- *A DJ's Guide to Latin Music*, by Jose Miguel, Chuck Fresh, Joseph Montalvo
- *Do What You Love, The Money Will Follow*, by Marsha Sinetar
- *First Things First; The 7 Habits of Highly Effective People*, by Stephen Covey
- *Guerilla Marketing*, by Jay Conrad Levinson; *Guerilla Marketing Attack*, by Jay Conrad Levinson
- *How to Avoid DJ Horror Stories*, by Jeff Harrison, David Westenbarger
- *How to Be a DJ: Your Guide to Becoming a Radio, Nightclub or Private Party DJ; The Ultimate Nightclub and Bar DJ Manual; Make Some Noise; Over 101 Great Things to Do at a Club, Bar or Party*, by Chuck Fresh
- *Innovation and Entrepreneurship*, by Peter F. Drucker
- *Lighting Dimensions International*, by Jane Hogan
- *Looking for the Perfect Beat: The Art and Culture of the DJ*, by Kurt B. Reighley
- *Making Money as a Mobile Entertainer*, by Raymond A. Mardo III
- Mobile Beat *Gear Book*
- *Real Magic: Creating Miracles in Everyday Life*, by Dr. Wayne W. Dyer
- *Sound Reinforcement Handbook*, by Gary Davis, Ralph Jones
- *Spinnin' 2000*, by Bob Linquist, Dennis Hampson
- *Success Through a Positive Mental Attitude*, by Napoleon Hill and W. Clement Stone
- *Testa Communications Blue Book*
- *Turning Music into Gold*, by Jeff Mulligan, Ryan Burger
- *The* Billboard *Book of Top 40 Hits* (6th ed.), by Joel Whitburn
- *The* Billboard *Book of Number One Hits* (1997), by Fred Bronson
- *The* Billboard *Book of Top 40 Country Hits*, by Joel Whitburn
- *The Green Book*, by Jeff Green
- *The Mobile DJ Handbook: How to Start and Run a Profitable Mobile Disc Jockey Service*, by Stacy Zemon
- *The One Minute Manager*, by Kenneth Blanchard and Spencer Johnson
- *The Power of Positive Thinking*, by Norman Vincent Peale
- *The Professional Guide to Coordinating Weddings*, by Henry Baker

- *The Seven Spiritual Laws of Success,* by Deepak Chopra
- *Think and Grow Rich,* by Napoleon Hill
- *Billboard* magazine—www.billboard.com
- *Dance Music Authority* magazine—www.dmadance.com
- *DJ Magazine* Online—www.djmag.com
- *DJ Times* magazine—www.djtimes.com
- *Djzone* Online magazine—www.djzone.net
- *Entrepreneur* magazine—www.entrepreneur.com
- *Hitmakers* magazine—www.hitmakers.com
- IRS Publications and Forms—www.irs.gov
- *Karaoke Singer* magazine—www.eatsleepmusic.com
- *Mobile Beat* magazine—www.mobilebeat.com
- *Nightclub & Bar* magazine—www.nightclub.com
- *Remix* magazine—www.remixmag.com
- *Singer* magazine—www.singermagazine.com
- *Spin* magazine—www.spin.com
- *The Music Yellow Pages Directory*—www.musicpages.com

Cases/Covers/Furniture

- American DJ—www.americandj.com
- Anvil—www.anvilcase.com
- Audioarts—www.wheatstone.com
- BGW—www.bgw.com
- Bryco Products—www.brycoproducts.com
- Cadence—www.cadencecases.com
- DJCases.com—www.djcases.com
- DJ Foundation—www.themobiledjhandbook.com
- (Formal) DJ Skirts International—www.djskirts.com
- Gem Sound—www.gemsound.com
- Gemini Sound Products—www.geminidj.com
- Grundorf Corporation—www.grundorf.com
- Horizon Music—www.horizonmusic.com
- Island Cases—www.islandcases.com
- Jewelsleeve—www.jewelsleeve.com
- MBT Lighting & Sound—www.mbtinternational.com
- Middle Atlantic Products, Inc.—www.middleatlantic.com
- Murphy Studio Furniture—www.murphystudiofurniture.com
- Nigel B. Furniture—www.nigelb.com

- Odyssey—www.odysseygear.com
- Omnirax—www.omnirax.com
- Pro Tec International—www.ptcases.com
- SKB Corporation—www.skbcases.com
- TopTone MFG—www.toptonemfg.com
- Univenture—www.univenture.com

Chat Sites

- www.djcafe.com
- www.djchat.com
- www.themobiledjhandbook.com
- www.prodj.com

Conferences/Competitions/Expos/Trade Shows

- *Billboard* Dance Music Summit—www.billboard.com
- Canadian Music Week—Conference Trade Show Exhibition of Awards—www.cmw.net
- DJ Cruise—www.djcruise.com
- *DJ Times* International DJ Expo—www.djtimes.com
- International Nightclub and Bar Convention & Trade Show—www.nightclub.com
- Karaoke Expo—www.karaokescene.com
- Lighting Dimensions International (LDI)—www.ldishow.com
- *Mobile Beat* DJ Show & Conference—www.mobilebeat.com
- NAMM—National Associations of Music Merchants—www.namm.com
- The Mid-America DJ Convention—www.midamericadj.com

Consultants

Ryan Burger—rburger@prodj.com
DJ Internet Consultation and Editing
(800) 257-7635

Ray Mardo—raymardo@raymardo.com
DJ Training Consultant

P.O. Box 328
Wayne, NJ 07474-0328
(973) 703-3568

J. R. Silva—funtalent@aol.com
DJ Training Consultant
(407) 426-9940

Sid Vanderpool—admin@djzone.com
DJ Internet Development Consultant

Stacy Zemon—djstacyz@aol.com
DJ Business, Marketing & Success Consultant
(413) 522-2090
www.themobiledjhandbook.com

Education/Schools

DJ Training Center and Studio
John Roberts
P.O. Box 565
Waldorf, MD 20601
(301) 843-6688
www.djtrainingcenter.com

Florida Academy of Mobile Entertainment DJ School
5611 65th Terrace N.
Pinellas Park, Fl 33781
(727) 531-8880
www.famedjschool.com/contact.html

The New York DJ Entertainment School
Frank Garcia
41-23 162nd Street
Flushing, NY 11358
(718) 359-4848

- Disc Jockey 101—www.discjockey101.com
- DJs Guide to Scratching—www.discjockey101.com
- DJ Foundation—www.themobiledjhandbook.com
- DJ Zone—www.djzone.com

- Instructional DJ Library on Cassette—www.mobilebeat.com
- ProDJ.com—www.dju.prodj.com
- *The DJs Guide to Running Weddings—www.proweddingguide.com*

Headphones/Microphones

- AKG—www.akg-acoustics.com
- Anchor Audio—www.anchoraudio.com
- Audio-Technica—www.audio-technica.com
- Audix—www.audixusa.com
- Azden Corporation—www.azdencorp.com
- BeyerDynamic—www.beyerdynamic.de
- Clear-Com—www.clear-com.com
- Countryman—www.countryman.com
- Furman—www.furmansound.com
- Gemini Sound Products—www.geminidj.com
- H&F Technologies—www.audio2000s.com
- HM Electronics—www.hme.com
- Hosa—www.hosatech.com
- Koss—www.koss.com
- Microtech-Gefell—www.microtechgefell.com
- Nady Systems, Inc.—www.nadywireless.com
- RTS Systems—www.telex.com
- Sabine—www.sabineinc.com
- Samson Technologies Corporation—www.samsontech.com
- Sennheiser—www.sennheiserusa.com
- Shure, Inc.—www.shure.com
- Sony—www.sel.sony.com
- Speco—www.csi-speco.com
- Tannoy—www.tannoy.com
- Telex—www.telex.com
- TOA Electronics—www.toaelectronics.com
- VocoPro—www.vocopro.com

Karaoke

- CAVS—www.cavsusa.com
- DB Karaoke—www.dbkaraoke.com
- H&F Technologies—www.audio2000s.com
- Karaokeholics—www.karaokeholics.com
- Music Maestro—www.musicmaestro.com

- Sound Choice—www.soundchoice.com
- VocoPro—www.vocopro.com

Lighting Equipment & Effects

- American DJ—www.americandj.com
- Bulbtronics—www.bulbtronics.com
- Carvin—www.carvin.com
- Chauvet Lighting—www.chauvetlighting.com
- Gem Sound—www.gemsound.com
- H&F Technologies—www.audio2000s.com
- Hosa Technology, Inc.—www.hosatech.com
- JM Electronics—www.jmelectronics-online.com
- Littlite—www.littlite.com
- Martin Professional—www.martin.dk
- MBT Lighting & Sound—www.mbtinternational.com
- NSI—www.nsicorp.com
- Pro Co—www.procosound.com
- TOV Lighting—www.tovlighting.com
- Tracoman—www.tracoman.com

Marketing

- Breakthrough Marketing—
  www.breakthroughbrochures.com
- Images Plus—www.imagesplus.com
- Night Club Items—www.nightclubitems.com
- Stacy Zemon—djstacyz@aol.com
- Stormline Media—www.stormline.com

Music/Culture

- Association For Wedding Professionals International—
  www.afwpi.com
- Basic.ch DJ mixes—www.basic.ch
- DigitalScratch—www.Digital Scratch.com
- DJ Approved—www.djapproved.com
- Disc Jockey 101—www.discjockey101.com
- DJ café—www.djcafe.com
- DJ Dot Net—www.dj.net

- DJ Foundation—www.themobiledjhandbook.com
- DJQBert—www.ses.djqbert.com
- DJ Zone—www.djzone.net
- E-Dancing—www.edancing.com
- Find Your DJ—www.findyourdj.com
- Hyperreal—www.hyperreal.org
- MTV Online—www.mtv.com
- Much Music—www.muchmusic.com
- MusicRemedy—www.musicremedy.com
- Pro DJ—www.prodj.com
- Pro Wedding Guide—www.proweddingguide.com
- Ray Mardo's DJ Tips—www.raymardo.com
- The DJ Society—www.djsociety.org
- VIP guestlist—www.vipguestlist.com

Music/Remix Services

- A Sound Image—www.asoundimage.com
- Bobby Morganstein Productions—www.powerhouse-remix.com/bmp_prod.html
- DJ Foundation—www.themobiledjhandbook.com
- DJ Music Express—www.djmx.com
- DJ Times' dance music—www.djtimes.com
- DJ Wholesale Club—www.djwholesaleclub.com
- ERG—Entertainment Resources Group—www.ergmusic.com
- ETV Network—www.etvnet.com
- Hot Tracks—www.hottracks.com
- Prime Cuts—www.productionmusic.com
- Promo Only—www.promoonly.com
- TM Century—www.tmcentury.com
- Top Hits U.S.A—www. tophitsusa.com
- Ultimix—www.ultimix.com
- X-MIX—www.xmix.com

Novelty Products/Party Props

Creative Imagineering—www.creativeimagineering.com

- Kabuki Confetti/Streamer Systems—www.kabuki.com
- Oriental Trading Company—www.orientaltrading.com

- PartyPoints Party Supplies—www.partypoints.com
- Pinto Novelty Co.—www.partypinto.com
- Sherman Specialty—www.makesparties.com
- Sure Glow—www.sureglow.com
- X-Streamers—www.x-streamers.com
- The Wedding Wheel/Party Wheel— www.weddingwheel.com
- Night Club Items—www.nightclubitems.com
- DJ Foundation—www.themobiledjhandbook.com

Organizations

United States

The American Disc Jockey Association (ADJA)
1964 Wagner Street
Pasadena, CA 91107
(626) 844-3204
www.adja.org

National Association of Mobile Entertainers (N.A.M.E.)
Box 144
Willow Grove, PA 19090
Phone: (215) 658-1193/1-800-434-8274 Fax: (215) 658-1194
www.djkj.com

- United States Online Disc Jockey Association— www.usodja.com
- DJ Foundation—www.themobiledjhandbook.com
- Association For Wedding Professionals International— www.afwpi.com

Canada

The Canadian Association of Mobile Entertainers and Operators
100 Blair Road
Cambridge, Ontario N1S 2J3
(519) 623-8719
E-mail: info@cameodj.com
www.cameodj.com

The Canadian Disc Jockey Association (CDJA)
P.O. Box 92
Arva, On Canada
N0M 1C0
1-877-472-0653 Toll Free
(519) 457-0742 55 Fax
www.cdja.org
Canadian Online Disc Jockey Association—www.codja.com

Software Products/Website Designers

- Dartech—www.dartpro.com
- DJ Calendar—www.djcalendar.com
- DJ Intelligence—www.djintelligence.com
- DJ ISP—www.djisp.com
- DJ Manager—www.djmgr.com
- DJ Power International—www.djpower.com
- DJ Soft—www.djsoftinc.com
- DJ Technology Alliance—
  www.mydjservice.com/djtechnologypackages.asp
- DJ Webmin—www.djwebmin.com
- DJ Zone—www.djzone.com
- Ebtech—www.cymation.com
- Event Electronics—www.event1.com
- Jampro—www.itc-net.com
- Mix Meister—www.mixmeister.com/mobile
- Music Database 2000—www.accsi.com/music/database.html
- ProDJ.com
- Real Networks—www.realnetworks.com
- Sonic Foundry—www.sonicfoundry.com
- SS7X7—www.ss7x7.com
- Stormline Media—www.stormlinemedia.com
- The Sonic Spot—www.sonicspot.com
- VisioSonic—www.visiosonic.com
- Visual DiscoMix—www.visualdiscomix-usa.com

Sound Equipment Manufacturers

(Amplifiers/Equalizers/Mixers/Processors/Speakers)

- Aardvark—www.aardvark-pro.com

- Akai—www.akai.com
- American DJ—www.americandj.com
- Aphex Systems—www.aphex.com
- Arrakis Systems—www.arrakis-systems.com
- ART—www.artroch.com
- Ashly—www.ashly.com
- Astatic—www.astatic.com
- Atlas-Soundolier—www.atlas-soundolier.com
- Audioarts—www.wheatstone.com
- BBE Sound, Inc.—www.bbesound.com
- B-52 Pro Audio—www.B-52pro.com
- Behringer—www.behringer.de
- Benchmark—www.benchmarkmedia.com
- Best Power—www.bestpower.com
- BGW—www.bgw.com
- Bogen—www.bogen.com
- BSS Audio—www.bss.co.uk
- Carver Professional—www.carverpro.com
- Carvin—www.carvin.com
- Celestion—www.celestion.com
- Clear-Com—www.clearcom.com
- Community—www.loudspeakers.net
- Crest Audio—www.crestaudio.com
- Crown International—www.crownaudio.com
- dbx—www.dbxpro.com
- Denon—www.denon.com
- DOD—www.dod.com
- Drawmer—www.drawmer.co.uk
- EAW—www.eaw.com
- Electrix—www.electrixpro.com
- Electro-Voice—www.electrovoice.com
- Eventide—www.tide1.eventide.com
- Fostex Corp. of America—www.fostex.com
- Furman—www.furmansound.com
- Galaxy—www.galaxyaudio.com
- Gemini Sound Products—www.geminidj.com
- Gem Sound—www.gemsound.com
- Gepco—www.gepco.com
- Hafler—www.hafler.com
- H&F Technologies—www.audio2000s.com
- HM Electronics—www.hme.com

- Horizon Music—www.horizonmusic.com
- JBL Professional—www.jblpro.com
- Juice Goose—www.juicegoose.com
- JVC—www.pro.jvc.com/prof/jvchome1.html
- Electro-Voice—www.electrovoice.com
- Klipsch—www.klipsch.com
- Korg—www.korg.com
- Lexicon—www.lexicon.com
- Mackie Designs—www.mackie.com
- Marantz—www.marantz.com
- Meyer Sound Labs—www.meyersound.com
- MicroAudio—www.microaudio.com
- MicroBoards—www.microboards.com
- Mobolazer—www.mobolazer.com
- MTX—www.mtxaudio.com
- Numark—www.numark.com
- Panasonic—www.panasonic.com
- Paso—www.pasosound.com
- Peavey Electronics Corporation—www.peavey.com
- Phonic—www.phonic.com
- Pioneer New Media Technologies—www.pioneerprodj.com
- PreSonus—www.presonus.com
- Pyle Industries—www.pyleaudio.com
- QSC Audio—www.qscaudio.com
- RANE—www.rane.com
- Renkus-Heinz—www.renkus-heinz.com
- Roland Corporation—www.rolandgroove.com
- RTS Systems—www.telex.com
- Samson Technologies Corporation—www.samsontech.com
- Shure, Inc.—www.shure.com
- Sony—www.sel.sony.com
- Sound Bridge—www.soundbridge.com
- Soundtech—www.soundtech.com
- Stanton—www.stantonmagnetics.com
- Tascam—www.tascam.com
- Teac—www.tascam.com/products/index.php
- Technomad—www.technomad.com
- Telex—www.telex.com
- TL Audio—www.tlaudio.co.uk
- TOA Electronics—www.toaelectronics.com
- TOV Lighting—www.tovlighting.com

- Trutone—www.trutone.com
- Vestax America—www.vestax.com
- Yamaha—www.yamaha.com/proaudio
- Yorkville Sound—www.yorkville.com

The author provides this Appendix for informational purposes only. She does not necessarily endorse and does not take responsibility for the listed manufacturers or their products. Apologies are made for any accidental errors or omissions.

Index